D0717977

The curious gardener's ALMANAC

The curious gardener's ALMANAC

centuries of practical garden wisdom

Niall Edworthy

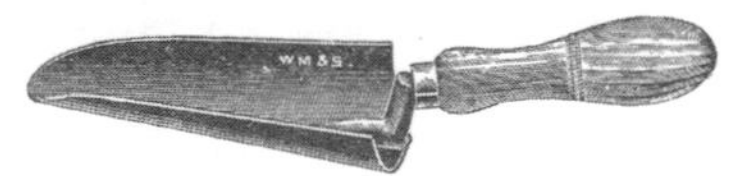

eden project books

TRANSWORLD PUBLISHERS
61-63 Uxbridge Road, London W5 5SA
a division of The Random House Group Ltd

RANDOM HOUSE AUSTRALIA (PTY) LTD
20 Alfred Street, Milsons Point, Sydney,
New South Wales 2061, Australia

RANDOM HOUSE NEW ZEALAND LTD
18 Poland Road, Glenfield, Auckland 10, New Zealand

RANDOM HOUSE SOUTH AFRICA (PTY) LTD
Isle of Houghton, Corner of Boundary Road and Carse O'Gowrie,
Houghton 2198, South Africa

Published 2006 by Eden Project Books
a division of Transworld Publishers

A catalogue record for this book is available
from the British Library.
ISBN 978-1-903919-90-3
EAN 9781903919903

Typeset in 10/12pt Mrs Eaves Roman by
Falcon Oast Graphic Art Ltd

Printed in Great Britain by Clays Ltd, St Ives plc
Design layouts by Fiona Andreanelli

13 15 17 19 20 18 16 14 12

This Almanac includes advice from historical sources; the author and publishers cannot be
held accountable for contradictions or anachronisms.

To my mum and dad,
Elizabeth and Patrick Edworthy

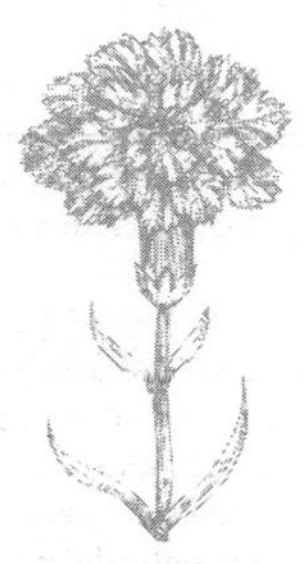

Acknowledgements

Some books are a hard slog to write or assemble, but this has been a rare pleasure and part of that is down to the people who have helped me along the way. The first person I need to thank is Susanna Wadeson, my excellent editor at Eden Project Books. She has been highly encouraging throughout and also very helpful in pointing me in the direction of the right people to talk to and the best places to visit for information. I owe a further debt of gratitude to copy-editor Deborah Adams and the design team at Transworld, who have come up with some very imaginative, original and clever work for a book that does not lend itself readily to straightforward design. (In fact, it must have been a bit of a nightmare.) Step forward Fiona Andreanelli, Lucy Davey, Sheila Lee, Phil Lord and Gavin Morris. Thanks also to Steve Apps, who, although he probably doesn't realize it, has taught me a good deal, and to Susan Maguire, an occasional lecturer at West Dean College and a walking encyclopaedia of organic gardening knowledge. Finally, ongoing thanks to my agent Araminta Whitley and her assistant Lizzie Jones at Lucas Alexander Whitley (LAW).

For permission to reprint lines from 'The Waste Land' by T.S. Eliot, my thanks to Faber & Faber Ltd; for lines from 'Spring Pools' by Robert Frost, from *The Poetry of Robert Frost* edited by Edward Connery Lathem and published by Jonathan Cape, my thanks to The Random House Group Ltd.

For the record, I would like also to pass on absolutely no thanks whatsoever to the cats Posh and Lenny and the dog Bruno, who continue to regard my vegetable patch and flower beds as a series of magnificent public conveniences laid on for their comfort and enjoyment.

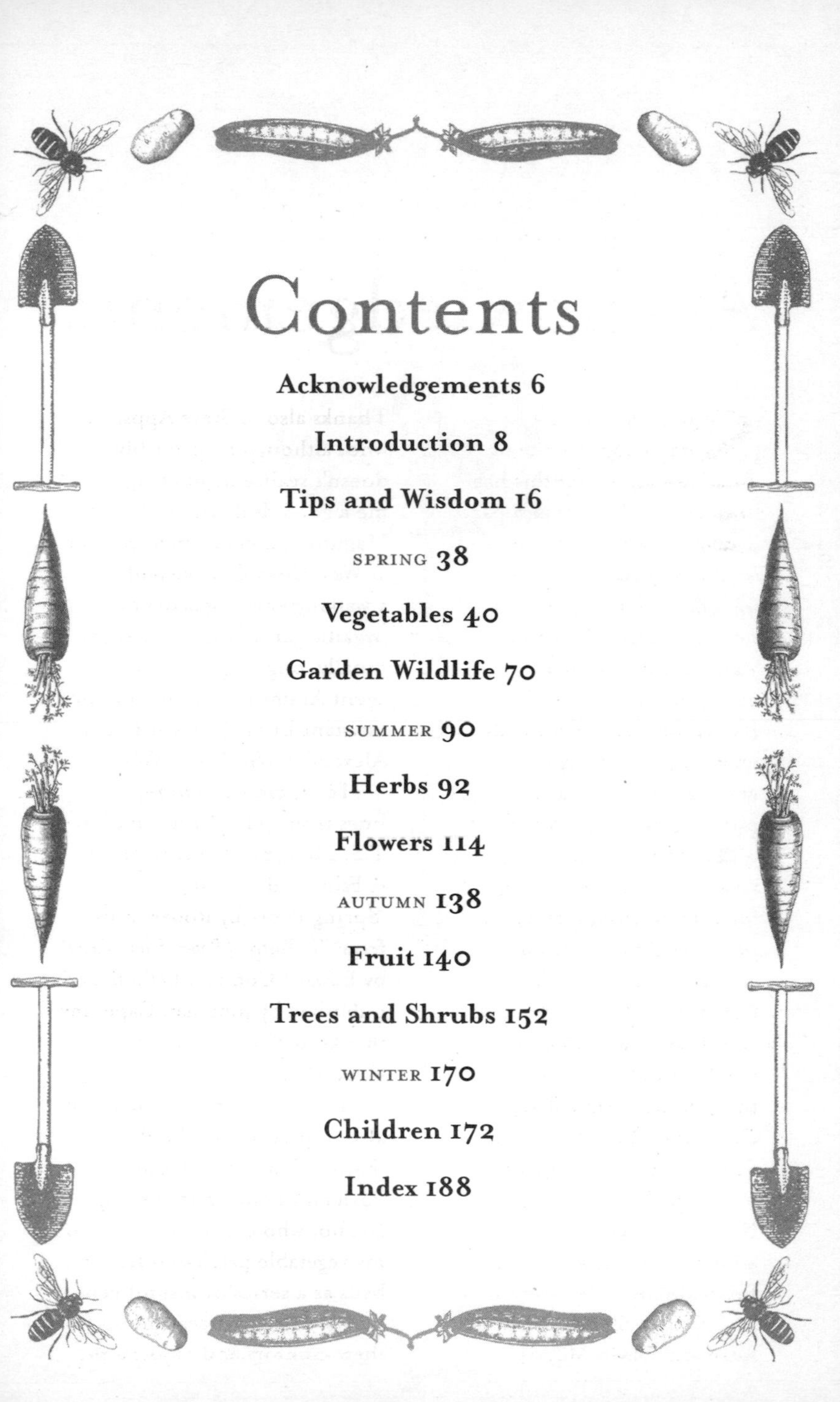

Contents

Introduction

IF THIS BOOK were to lie down on a couch and open up to reveal everything that was going on deep inside, the psychiatrist would probably get up and walk out of the room, pulling at his hair and kicking over the wastepaper basket in exasperation as he left. *The Curious Gardener's Almanac* has an identity crisis of the first order. Is it a book of time-honoured garden tips and wisdom? A compendium of weird facts and ephemera about nature? A call to gardening and environmental arms? An indispensable resource for the setter of pub quizzes? A meandering narrative with a few jokes and recipes, a bit of homespun practical advice, a smattering of sayings and some poignant or witty verse?

Its natural habitat in the bookshop is the Gardening shelves, but it could also slip into Nature or Natural History without raising too many eyebrows and, at a squeeze, it might not be completely out of place in Cookery, Environment, Children or Humour. (Fingers crossed it doesn't end up in Hobbies and Pastimes or down on the bottom shelf in Cobweb Corner sandwiched between *Two Hundred Years of Estonian Free Verse* and *The History of the Italian Aviation Industry*.)

Reading through the text a final time, I confess that I am no closer to being able to describe exactly what it is that (I hope) you are about to read. When people asked me, while I was putting this book together, what I was working on, I replied, 'Um . . .' and then, lost in the muddle of my own thoughts, quickly changed the subject to Gordon Brown's chins or Accrington Stanley. The word 'miscellany' keeps popping into my head, only to be promptly thrown out on its ear, partly because the book isn't simply a collection of odd facts, but mainly because when I say it, 'miscellany' sounds as if I'm choking on a mouthful of mashed potato.

Luckily for me, my publishers had the good grace to pretend they knew what I had in mind, i.e. a big mishmash of anything to do with gardens that I found interesting, weird, funny, useful, charming, surprising or delicious (see recipes). The only guiding principle was that every entry, of which there are roughly a thousand, had to be able to attach itself to at least two of those adjectives, without running the risk of me being spat at in the street by Alan Titchmarsh or anyone else who happens to know far more about these matters than I do (i.e. roughly 30 million Britons). Dull information was to be turned away at the door before it had a chance to brush its feet and reach for the bell.

There have been plenty of excellent gardening books, and a few dodgy ones too, that have set out with a single objective – 'How To' manuals, books of garden verse, sayings, tips, recipes and so on – but this book has tried to take the best of all those worlds, sifting the wheat from each field and dispensing with the chaff. The *Almanac* is not a book that can be read – or, indeed, wants to be read – in a single

sitting or over long periods. It is presented so that readers may delve into it from time to time in the hope that they will be intrigued, amused, enlightened, surprised, or even inspired. It could do worse than set up home in your loo, but you may also want it in the kitchen or in the shed or the greenhouse, or perhaps on your bedside table. It may also be of interest to your children, for whom a whole chapter has been laid out, and they may well enjoy the other chapters too because I have filled them with as many eye-widening, jaw-dropping facts as possible.

If you were to put a gun to my head and demand I describe this book to you in just a few words, there is a small chance I would wrestle you to the ground, put you in a half-nelson and call for police back-up. It's more likely, however, that I would get down on my knees and whimper in a high-pitched voice: 'It's a celebration of nature, in all its infinite variety, as experienced in the British garden . . .' at which point you could put the gun down and we could carry on as if everything was completely normal.

I didn't set out to make this book 'a celebration of nature', but with hindsight I now wonder how it could ever have been anything else. Not through any conscious effort on my part, the book has ended up with an environmentalist message. It is not, by any stretch of definitions, a tub-thumping, prescriptive treatise, or a rallying cry to the masses to rush out and hang our farmers, polluters and politicians, but any book that invites you to relish the wonder, charm and glory of the natural world sitting on our doorsteps cannot help but become, by implication, a mournful commentary on the enormous degradation that our environment has suffered in the past 50 years.

The populations of many birds, animals, insects and reptiles have plummeted in this time, while many ponds, woods, marshes, waterways, meadows, plants, hedgerows and flowers have disappeared or are under threat as a result of modern farming practices, industrial and general pollution, climate change, the growth of towns and the burgeoning infrastructure demanded by our ever-expanding communities. Regiments of foot-soldier gardeners, however, are joining the cavalry of environmentalists in coming to the rescue, whether they realize it or not. Gardens make up roughly 15 per cent of the land area in British towns and cities, while 10 million of us are said to be keen gardeners, and a further 20 million could become so, should the wish ever take them. Gardeners, therefore, have the opportunity to make a significant contribution to preserving and boosting our natural heritage. By planting a few common plants, hedges, shrubs or trees for their own enjoyment and benefit, gardeners also unwittingly encourage the wildlife that needs those plants as a habitat or a food source. By using natural, non-chemical methods that worked perfectly well for our ancestors for century upon century, the gardener will also be helping to keep his environment in the pristine manner to which it had become accustomed until roughly half a century ago. And this simple, innocent way of gardening is not rocket science.

Just a few years ago I could never have imagined that I would write a book about gardening. As a ghostwriter I have written autobiographies for actors, sportsmen and television personalities, as well as, under a variety of names, humour books, outdoor survival guides, TV tie-ins and a military history. But gardening – you had to be kidding! That was what my mum and dad did because, for some extraordinary reason, probably generational, they had yet to discover the more rewarding pastimes of sitting inside watching the cricket or going to the pub.

Then something weird happened. On 17 March 2003, to be precise. After almost 20 years of living in garden-less properties in Edinburgh and London, I moved to West Sussex, with my wife and two young kids, in search of a slightly larger house and (we hoped) a life with a little less stress and a little more time to enjoy ourselves. We found a house in a village that ticked most of the boxes (cricket green, good pub), and I couldn't help but notice as we unpacked our last tea crates and suitcases that our new home was surrounded by a green area filled with flowers, shrubs, a few trees and hedges, a greenhouse, a shed and something I dimly recognized from the television as a vegetable patch. I took little notice of it all at the time, except to make a mental note that this strange place seemed, on first impressions, to be a superior form of environment to the concrete and burning motorbikes we had enjoyed in our corner of Hammersmith. It would be excellent too for letting our children run around in, I thought.

A month or so later, I glanced out of the window and

noticed that this green area had turned into something of a jungle. Blimey, this stuff grows. The grass, as they called it in the country, had become shaggy, weeds were starting to battle the flowers and other plants for space, and the vegetable patch, once brown with mud, was now a carpet of green. It was in that one moment that I realized I had no choice but to become a gardener. I would love to tell you that a strange light then suffused the horizon and St Fiacre, the Irish monk and patron saint of gardeners, appeared over the hedge and began to hover over the vegetable patch, beckoning me towards him. In truth, I groaned inwardly. Gardening had just joined the ever-growing list of 'Things to Do Today' that I like to write out every morning and then ignore for the rest of the day. Hiring a gardener for more than an hour or two a week was out of the question on financial grounds because I had had

to give up my day job as a journalist when we left town. There was only one thing for it – my wife Charlie would have to do it all, ably assisted by my one-year-old daughter Eliza and three-year-old son Alfie. Just kidding – they had quite enough on their plates as it was. There was only going to be one steward of this plot and that was me. So with a heavy heart I climbed into my car and headed off to the local garden centre to arm myself with Dutch hoes, dibbers, hand sporks, rakes, etc . . .

The first year in the vegetable patch was a perfect disaster – I just scattered a variety of seeds over it, expecting it to turn into the Garden of Eden by the end of summer, like it does on the telly. The few sprigs and sprouts that did make it out of the ground reached no higher than about two or three inches before some form of wildlife gratefully snapped them up. What form these predatory beasts took, I had simply no idea. Mice? Slugs? Deer? Birds? Dolphins? Thanks entirely to local gardener Steve Apps, who came in for two hours a week (and still does, thank God), the rest of the garden was at least under control and looked pretty good, considering I could ask him to drop by for a total of only eight hours a month. The following year, during which I bought a couple of books and tortured some locals for advice, there was a slight improvement. I could now name about four flowers and, incredibly, I even managed to produce a few herbs as well as some potatoes, tomatoes, spinach and one very sad, small aubergine that was so hard I used it as a paperweight. As I reflected on my pathetic efforts that autumn, I had to face the fact that, 18 months on, I knew precisely sweet bugger-all about gardening.

If there is one factor, above all others, that will motivate a man with a garden to become a better gardener it is the embarrassment of walking round it with a friend, relative or local acquaintance who happens to be pretty handy with a trowel and a watering can. They look around sadly and, being well brung up, say nothing to injure your feelings. At worst, they let out a barely audible sigh of disappointment. It's what interrogators call the silent treatment and eventually I could take no more of it. I resolved that by spring I would have a good working knowledge of the basics of gardening. I enrolled on a two-day course at West Dean College (home to some well-known, beautiful gardens and some superb courses), just over the hill from where we were living; I bought half a dozen recommended books and committed myself to badgering to distraction – and even to death by boredom – anyone with even the slightest amount of garden experience and knowledge. Many good people died.

What I came to discover, in between burying the neighbours, was that, while there is no end to how much there is to learn about gardening, the basics are just that: basic. Even the dribbling village idiot (that's me) could grasp the fundamental knowledge needed to create a perfectly happy, half-decent garden that looks OK and yields a steady stream of produce for the kitchen table – and I realized that once you knew these basics, you didn't need to 'do a Lucan' and disappear into the shed, never to be seen again, in order to achieve this. I also discovered that you don't grow a good garden by reading books alone; you have to get down and dirty and learn by trial and error.

At this point I am tempted to tear off my literary mask and announce to you that I am, in fact, the love child of Alan Titchmarsh and Vita Sackville-West; that my gardens are the most beautiful in the temperate zone of the northern hemisphere and that my vegetable patch could feed the entire Chinese army and still have some left over for my own kitchen table. But these would be tremendous lies, and I mustn't get carried away.

The prosaic truth is that my garden now looks quite a bit better than it did. It has more wildlife in it than before, in spite of the slaughter our two cats have visited upon the local mouse and vole population. In 2005 the vegetable patch and greenhouse were successful enough to yield roughly three or four months' worth of perfectly acceptable fresh, healthy produce. We were even able to put a few bags in the freezer and fill some jars with funny pickled things. For the sake of accuracy, I have to confess also that, quite amazingly, I did manage to bugger up the onions (which really are not very difficult to grow) by leaving them out to dry for too long. On Day Seven in the rain, they succumbed to a rotten death.

So that, in a very roundabout fashion, is how this book came into existence. In short, I

inherited a garden. I didn't have the first clue what to do in it, my neighbours openly wept with laughter at my Laurel and Hardy efforts, I set out on a quest for knowledge, I prevailed over the slugs, aphids and mice and I can now stand proudly over my little plot of land and say, with a slight burr, to my new gardening chums, 'The asparagus is up late this year, isn't it?' To which they reply, shuffling from foot to foot: 'Er, I think you'll find those are the potatoes.'

Tips and Wisdom

Everybody talks about the weather but nobody does anything about it

Some of the most delightful of all gardens are the little strips in front of roadside cottages. They have a simple and tender charm that one may look for in vain in gardens of greater pretension.

GERTRUDE JEKYLL (1843–1932)
gardener and pioneer of the quintessentially English informal garden

The more one gardens,
the more one learns;
and the more one learns,
the more one realizes
how little one knows.
I suppose the whole of
life is like that.

VITA SACKVILLE-WEST
(1892–1962)

PLINY THE ELDER'S *Naturalis Historia*, a 37-volume encyclopaedia about the natural sciences, was widely read by herbalists, gardeners and naturalists throughout the Middle and Tudor Ages. He died during the eruption of **Vesuvius** in AD 79, when, as commander of the Roman fleet at Misenum, he sailed close to the shore to investigate and was promptly overwhelmed by the fumes.

THE BEST TIME TO **TRANSPLANT trees and shrubs** is spring and early autumn so that the roots have time to establish themselves before the heat of the summer or the cold of the winter.

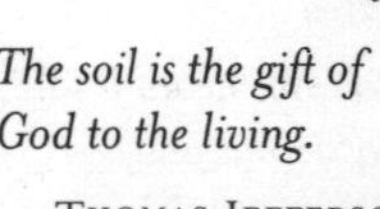

The soil is the gift of God to the living.

THOMAS JEFFERSON
1813

The watering of a garden requires as much judgement as the seasoning of a soup.

HELENA RUTHERFORD ELY
A Woman's Hardy Garden, 1903

CUTTING AN UNRIPE TOMATO IN half and rubbing the juice on your fingers before washing will remove **stubborn green stains**. Other handy tips for cleaning hands include mixing a teaspoon of sugar with the lather from your soap and rubbing your hands with the cut end of a rhubarb stick.

A dunghill at a distance smells like musk, and a dead dog like elderflowers.

SAMUEL TAYLOR COLERIDGE
(1772–1834)

WATER YOUR GARDEN OR LAWN in the early evening or, failing that, in the early morning. Much of the moisture will be lost to evaporation if you water in the heat of the day, and on very hot days the leaves of some very tender plants may even 'burn'. Only water when your plants need it and make sure you water them well to encourage deeper root growth, rather than little and often, which will only encourage the roots to head towards the surface in search of moisture.

IF YOU THINK CIGARETTES MAY BE bad for *your* health, you should see what they do to the aphids chewing their way through your cherished plants. An easy-to-make tobacco solution is an effective and harmless – except to the **aphids and caterpillars** – method of controlling these pests. ('Controlling', of course, is a Downing Street Press Office way of saying 'brutally exterminating'.) Put a handful of rolling or pipe tobacco into a gallon of water, give it a stir and let it sit for 24 hours, then dilute it until it is a pale brown colour. Spray the solution on to the affected plants, and those aphids will wish they had never been born. (NB: Do not spray on potatoes, tomatoes, peppers, aubergines or any other members of the *Solanaceae* family, as the spray will kill them too.)

What is a weed? I have heard it said that there are sixty definitions. For me, a weed is a plant out of place.

DONALD CULROSS PEATTIE
(1896–1964)
botanist and author

In dry weather observe to water all such plants as have been lately transplanted, and to be sure always to do this in an evening, for one watering at that time, is of more service than three at any other time of the day, the moisture having time to penetrate the ground (and reach to the extream fibres of the root by which they receive their nourishment) before the sun appears to exhale it; whereas when it is given in a morning, the sun coming on soon after, the moisture is drawn up before it reaches the root.

PHILIP MILLER
The Gardeners Kalendar, 1732

IT IS SOMETHING OF AN URBAN myth that ivy will ruin the walls of a house. What is true is that if your walls have cracks, the ivy will find its way into them and may make them worse.

I have often thought that if heaven had given me choice of my position and calling, it should have been on a rich spot of earth, well watered, and near a good market for the production of a garden . . . though an old man, I am but a young gardener.

THOMAS JEFFERSON (1766–1824)
Garden Book

THE INVENTION OF the **wheelbarrow** is usually traced to China's Chuko Liang, an adviser to the Shu-Han Dynasty from AD 197 to 234, who had it developed as a means of transport for military supplies. The first evidence of wheelbarrows being used in Europe is found in illustrations in the thirteenth century.

Man was lost and saved in a garden.

PASCAL (1623–62)
Pensées

The taste of the English in the cultivation of the land, and in what is called landscape gardening, is unrivalled. They have studied nature intently, and discover an exquisite sense of her beautiful forms and harmonising combinations.

WASHINGTON IRVING (1783–1859)
The Sketch Book, 'Rural Life in England'

There is an old saying that if a shrub flowers before the middle of summer, prune it in the autumn; if afterwards, then wait till spring. There are exceptions to this, but generally it holds true.

When the world wearies and society fails to satisfy, there is always the garden.

MINNIE AUMONIER
(N.D.)

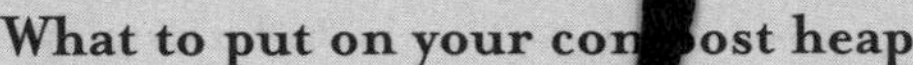

What to put on your compost heap

Old plants	Cardboard egg containers	Teabags, coffee grounds
Hedge trimmings	Shredded paper (not glossy mags or leaflets)	Bark
Fallen leaves	Natural fibres like cotton and wool	Roots
Grass cuttings		Nettles
Vegetable and fruit peel		Urine (human)
Eggshells		

Making compost

There are six essential ingredients for making good compost, all of which are free: organic matter, moisture, heat, air, insects/micro-organisms and time. For the perfect mixture you'll need a combination of greenery (lawn cuttings, leaves, flowers, vegetable and fruit peelings) and some woodier matter (stems, roots, hedge trimmings). Don't put leftover food into the mix unless you like rats. It is important that your compost bin has some open pockets through which insects can get in to help break it all down. Make sure you use a variety of ingredients and turn it all over with a garden fork from time to time to let the air in and so that the matter on the outside gets to enjoy the heat in the middle too. If your first compost heap includes a lot of lawn cuttings, which are high in nitrogen, then you probably won't need to add an 'activator' to get it going. If not, have a good wee on it whenever you can, or add a few bunches of nettles.

What NOT to put on your compost heap

Very woody clippings or branches	Bones
Manmade fabrics	Soil pests
Eggs	Weeds with seeds
Leftover food scraps, meat	Oil
	Diseased plant material

The labourer who possesses and delights in the garden appended to his cottage is generally among the most decent of his class; he is seldom a frequenter of the alehouse.

GEORGE W. JOHNSON
A History of English Gardening, 1829

MANY GARDENERS (MAINLY THE male of the species, it has to be said) swear by the **benefits of beer** as a plant fertilizer. It is thought to be good for root growth. Use bitter, not lager, and the more organic it is the more effective it will be, they say. The only creatures on the planet that seem to like beer more than men are slugs. Like men, slugs will often head out under the cover of darkness to find beer, but whereas men have developed an incredible ability not to fall into their pints, slugs have yet to reach that crucial stage in their evolution. Unfortunately for them, but happily for the gardener, the slugs will die when they drop into the submerged glass of ale you have generously left out for them.

It will be a key to right thinking about gardens if you consider in what kind of places a garden is most desired. In a very beautiful country, especially if it be mountainous, we can do without it well enough; whereas in a flat and dull country we crave after it, and there it is often the very making of the homestead. While in great towns, gardens, both private and public, are positive necessities if the citizens are to live reasonable and healthy lives in body and mind.

WILLIAM MORRIS
lecture entitled 'Making the Best of It', 1879

SLUGS detest the smell of peppermint, spearmint and elder, so an infusion made from the leaves of any of them, poured or sprayed around the plants you want to protect, will send them oozing and sliming their way to more attractive pastures. It is best to do this at dusk when the slugs start to emerge. **Rats** and other rodents, meanwhile, loathe the smell of tar, so if your garden has a problem with them (i.e. if you live in a city), soak some rags in tar and leave them scattered around. **Rabbits** hate foxglove and onions and it is unlikely they will bother your patch if you have good quantities of them in situ. **Mice** don't like the smell of mint.

The mouth of a perfectly happy man is filled with beer.

Ancient Egyptian saying

Think twice about using most types of **slug repellent** as dogs, cats and even small children are at risk from ingesting them. Hedgehogs are probably the best form of slug killer you could hope to find, but they will die if they eat slugs that have consumed toxic pellets.

Some older gardeners may recall the days when they were sent out into the street by their parents to collect piles of fresh steaming horse manure for the compost heap. **Compost**, as experienced gardeners never tire of telling their younger, more amateurish counterparts, is everything in the garden. 'Look after the soil, and the soil will look after the plants' . . . 'Feed the soil, not the plants' . . . There are many different types of compost and mulch you can produce yourself, but first you will need somewhere to pile it. You can buy plastic compost bins fairly cheaply these days, but if you prefer a more natural-looking structure, making your own bin is easy and cheap. Take a roll of tightly meshed chicken wire and tie four, six or eight posts (no more than the width of a cricket stump) at equal distances from each other and then push the sharp ends of the posts into the ground. Compost likes heat to break down all the material, so cut a piece of old carpet to size and place it on the top. This will also stop the rain from washing away some of the goodness. Animal manure should be stacked high and turned over regularly to stop it from overheating. If you can build your compost heap on stones or wire meshing so that it is slightly above the ground, then so much the better as this will help aeration.

An addiction to gardening is not all bad when you consider all the other choices in life.

Cora Lea Bell (n.d.)

TO KILL OFF **UNWANTED INSECTS** in the greenhouse or cold frame, bash up some **laurel leaves** and leave them in a bowl overnight. The damaged leaves exude prussic acid gas, which is deadly to most small insects.

THERE ARE A NUMBER OF CHEAP, painless ways of ridding your pathways or drive of **stubborn weeds** without resort to chemicals. A mixture of salt and ashes works well but even just pouring a watering can of hot salty water should do the trick.

Move the bird around every week or so, as the birds sitting up in the branches will eventually work out that you are nothing but a cheap, low-down fraudster. You could also make use of that wooden parrot mobile that used to hang above your toddler's cot but which has long been consigned to the bottom of the toy box. If you can bear the sight of them, those garish, colourful windmill flowers, beloved of seaside bucket-and-spade shops, are as unappealing to birds as they are to you.

> *If you want to be happy for a day, get drunk*
> *For a week, kill a pig*
> *For a month, get married*
> *For life, be a gardener.*
>
> CHINESE PROVERB

IF YOU HAVE TROUBLE WITH **birds** eating seeds and buds and uprooting young seedlings, a fake bird of prey often works as an effective deterrent (see also p. 69). Nor do you have to be an experienced *Blue Peter* presenter to make one: just stick some large feathers into a **large potato** and hang it over or close to your seed bed or vegetable plot. The bigger you make the wings and the more colourful the artwork of the 'feathers', the more effective will be the deterrent.

IT IS A **REGRETTABLE TRUTH** that the better the quality of the soil, the bigger and more prolific the weeds.

SOME GARDENERS WAX SO LYRICAL about the benefits to a garden of **Russian comfrey** you'd think they had come up with a cure for cancer or brought peace to the Middle East. It certainly is an effective mulch for a number of fruit bushes, flowers and vegetables, and even better when made into liquid fertilizer, packed with nitrogen and therefore superb for encouraging the leaf growth of your spinach and chard. Like wormcasts, Russian comfrey also contains high levels of potassium and phosphorus, and tests have shown it to be a superior compost to farm manure, which is some achievement. Leaves left to wilt for a day or two help the growth of potatoes if buried in the trench prior to laying your tubers. The liquid is also good for tomatoes and peppers once the flowers have set. If you apply it before then, you will only encourage lots more leaves at the expense of the flowers/fruits. ('Bocking 14' is the cultivar you want and it is very easy to get hold of.)

Russian comfrey is also a vigorous grower, and you can harvest it several times a season. Another of its benefits is that it tolerates a semi-shady position and looks quite attractive, especially when it flowers, so it makes good ground cover in the darker, sadder corners of your garden and will produce a good supply for up to 25 years. It will also attract bumble bees.

To make the liquid, all you need is a barrel with a lid on top and a tap close to the bottom. You can use an **old plastic dustbin** or rain butt, and then drill a hole near the bottom and insert a tap bought from a garden centre or ironmonger's. Fill the container with comfrey leaves and place something heavy on top of them to create some downward pressure – a flat piece of wood with a couple of bricks would be ideal. The barrel should be raised on bricks so that you can collect the liquid. Put the lid back on to stop flies getting in, and after about ten days the liquid will start to emerge. It's pretty powerful stuff, however, so be sure to dilute it well – roughly 15 parts water to one of the comfrey – or your plants will not thank you for the helping hand. The liquid will last for months if you store it somewhere cool. If comfrey liquid has one shortcoming, it is this: it absolutely stinks.

Cut a thistle in May, it will be back the next day,
Cut a thistle in June, it will be back soon,
Cut a thistle in July, it will surely die.

ENGLISH SAYING

STORE SEEDS SOMEWHERE COOL like the fridge or an old biscuit tin kept in the shed or garage.

This rule in gardening ne'er forget,
To sow dry and set wet.

ENGLISH SAYING

WHEN PLANTING SEEDS, MAKE sure the soil is not too loose. **Firm the soil down** by treading upon it after you have prepared the bed. A seed likes to know it's on reasonably firm ground when its first tiny root emerges.

A GOOD WAY OF USING UP old cardboard **egg cartons** in a manner that might be called direct recycling is to plant up your seedlings from the greenhouse in them as if they were little peat pots. You can plant them straight into the ground as they are biodegradable.

Herein were the olde husbands very careful and used always to judge that where they found the Garden out of order, the wife of the house (for unto her belonged the charge thereof) was no good huswyfe.

BARNABY GOOGE, 1390

THE FIRST GARDEN **GNOMES** were introduced to the United Kingdom in 1847 by Sir Charles Isham, who brought back 21 terracotta figures from Germany and placed them as ornaments in the gardens of his home, Lamport Hall in Northamptonshire. Only one of the original batch of gnomes survives: Lampy, as he is known, is on display at Lamport Hall, and is insured for one million pounds. Garden gnomes are not everyone's idea of artistic beauty in the garden, but some see them as an oppressed minority whose civil rights have been violated by unscrupulous landlords. In France there is an underground movement known as the Liberation Front of Garden Gnomes (Front de Libération des Nains de Jardin), while there is an equally shadowy organization in Italy known by the acronym MALAG. For years, these activists have been emerging under the cover of night to remove hundreds of gnomes from gardens and place them in prominent positions in city centres to bring their plight to the attention of the wider public. Some are relocated to the woods, where the **suburban terrorists** feel they belong – so, in the words of the French group, 'de-ridiculizing the figurines by placing them back in their natural environment.' In these more liberal times, it is now possible to buy garden gnomes that have been sculpted or moulded into scenes depicting them having sex or exposing themselves.

My garden will never
make me famous
I'm a horticultural ignoramus.

Ogden Nash (1902–71)
US poet

Tiny seedlings are easily damaged and you have to be careful when transplanting them. It's better to hold them by the leaves than the stem, for the simple reason that the stem is vital to the plant's growth as water and nutrients pass through its 'plumbing' system to reach other parts of the plant. A plant can always grow more leaves. Using a teaspoon to transport a seedling will minimize damage to the plant's delicate young structure.

Try to keep the leaves of your **seedlings clean**, and not covered in soil as they often become when transplanted. This is because the young plant needs to absorb as much light as possible through its small leaves in order to grow.

In days gone by, it was not uncommon in springtime to see a farmer drop his trousers and pants and **sit down on the soil**. If the soil was not too cold, the farmer knew it was time to sow his crop. If you want to avoid misunderstandings with your neighbours, you are better off trying this old trick with your bare elbow.

The reason why gardeners break off the **top shoot** of some plants, such as tomatoes, is to encourage more vigorous growth in the lower branches.

Four seeds you have to grow
One for the pheasant, one for the crow,
One to rot and one to grow.

Old saying

Jane Loudon, in *Gardening for Ladies* (1840), on the subject of **Victorian women** digging their gardens: *She will not only have the satisfaction of seeing the garden being created, as it were, by her own hands, but she will find her health and spirits wonderfully improved by the exercise, and by the reviving smell of the fresh earth.*

Only fools will lend their tools.

ANON

A COMPREHENSIVE SURVEY known as **BUGS**, recently carried out by Sheffield University and recorded in an excellent book called *No Nettles Required*, showed that when it comes to attracting wildlife to your garden, whether in the form of birds, animals or insects, it doesn't really matter whether the plants you choose are 'native', naturalized or imported. But if you do want to add new plants to your garden and are unsure what species and cultivars to go for, ask local gardeners or observe the plants that have been growing happily in your area of the country. The reason for this is self-evident: those plants obviously like the local soil, temperature, insect life and general environment. The Natural

History Museum provides a fabulous service on its website www.nhm.ac.uk, which invites you to input your postcode to find the plants best suited to where you live.

When all is said and done, is there any more wonderful sight, any moment when man's reason is nearer to some sort of contact with the nature of the world than the sowing of seeds, the planting of cuttings, the transplanting of shrubs or the grafting of slips.

ST AUGUSTINE (D. 604)

If your garden has a tendency to dry out quickly in warm weather, it's good sense to plant flowers and shrubs that enjoy those conditions. 'Mediterranean' plants like lavender and rosemary, as well as buddleia, sedum and sunflowers, can go for weeks on end without water.

Clay soils pose the opposite problem to sandy ones because they are so compacted and sticky that the ground cannot 'breathe' or drain effectively. Wet ground will also stunt root growth, or even rot the roots. Add gritty, pebbly sand (not the soft type used by builders and in sandpits) to the soil to improve the drainage and air circulation.

It's often difficult for **novice** gardeners to work out exactly what type of soil they have in their garden. To the inexpert eye, most earth just looks brown and dirty. Ideally, you want your soil to be a roughly equal balance of clay, sand and organic matter. You can determine the nature of your earth, and enjoy a trip back to your childhood, by performing a **simple experiment**. Fill about a third of a clear bottle with some soil, but make sure it comes from at least one foot down as the topsoil may have had compost and other matter heaped on it over the years. Pour in another third or so of water and give the bottle a good shake before leaving it to settle. The heavier sand will sink to the bottom pretty quickly and the organic matter will float to the top, while the finer sands and clay will take much longer to settle. If you check the bottle a few hours later, you will get a good idea of what makes up your soil, and you can act accordingly to correct the balance.

Plants are generally quite robust and adaptable, and a certain amount of dry weather may even be good for them because it will **encourage roots** to grow deeper in search of moisture, thereby making them stronger in the long term.

The fair-weather gardener, who will do nothing except when the wind and weather and everything else are favourable, is never master of his craft.

Canon Henry Ellacombe (1822–1916)

We come from the earth, we return to the earth, and in between we garden.

Anon

THE FURTHER SOUTH YOU LIVE in Britain, the greater the urgency to conserve as much water as possible. Summers are getting hotter and drier, and water authorities are starting to fret about supplies. Collect as much **rainwater** as you can by using water butts attached to the house, greenhouse, shed or other outbuildings. One of the added benefits of collecting your own water is that it is better quality than the 'dead water' that comes straight from the tap.

A man should never plant a garden larger than his wife can take care of.

T. H. EVERETT (N.D.)

SANDY SOILS DRAIN QUICKLY and therefore cannot hold on to moisture and nutrients long enough for the plant to benefit from them. Digging in organic matter will help bulk up the soil, allowing it to retain water and goodness for a lot longer. 'Organic matter' includes garden compost, used compost from pots, grow-bags, tubs and hanging baskets, and well-rotted manure. The use of peat is distinctly un-PC these days as its extraction wreaks havoc in ancient natural habitats, and you can expect to be button-holed by environmentalists if you are spotted creeping out of the garden centre with a bag of it on your trolley.

IF YOUR **HANGING BASKETS** HAVE a tendency to leak heavily after watering there are a number of ways of preventing this. Before adding the soil, place absorbent material at the bottom. Sphagnum moss, popular with our prehistoric ancestors (and which you can buy from garden centres), is capable of holding up to 20 times its weight in water. In fact the nomads of Stone Age Britain stored water in it when travelling from one site to the next. They also used it as nappies for their babies. Some hanging basket gardeners use modern-day nappies, or old woollen jerseys cut to size, to retain the water in the soil. Old newspapers will work for a while but will eventually decompose. If you haven't lined the basket before planting, water the plants with ice cubes, which will be absorbed slowly.

There is not a spot of ground, however arid, bare, or ugly, that cannot be tamed.

Gertrude Jekyll (1843–1932)

Loamy soil (i.e. that which is neither sandy nor clay) is best for the gardener because it drains well while retaining nutrients and water and there is enough air space for healthy root growth.

On the English: *Gardening is their greatest art. It is immensely widely spread, the interest in gardening and flowers. It is the most living art in this country, I think, and has been for a long time.*

Ernst Gombrich (1909–2001)
Austrian-born art historian

If you want to make a nice straight seed trench before planting, use an **old brush handle**. Lay it flat and tread it into the earth. (A similarly simple method applies to edging up your straight flower beds: lay a plank of wood on the grass and then press your spade down over the uneven patches protruding into the beds.)

When you come **back from holiday** and your grass is overgrown, be sure not to chop it right down to the required height but to cut it in two sessions with a few days in between. Drastic cutting upsets the grass and can affect the regrowth.

I suppose it is the same with everything in life that one really cares about, and you must not, any of you, be surprised if you have moments in your gardening life of such profound depression and disappointment that you will almost wish you had been content to leave everything alone and have no garden at all.

C. W. Earle ('Mrs Earle')
Pot Pourri from a Surrey Garden, 1897

A well-watered **lawn** will look great for a week or two, but you will have created a rod for your own back because it will start to expect another dousing. Grass is a tough plant and can survive very long stretches without water. No matter how severe a summer drought, no one's lawn in Britain has ever died from lack of water. It can survive up to eight months without rain.

I have read much and found nothing but uncertainty, lies and fanaticism. I know about as much today of the essential things as I knew as an infant. I prefer to plant, to sow, and to be free.

Voltaire (1694–1778)

God Almighty first planted a garden. And indeed it is the purest of human pleasures. It is the greatest refreshment to the spirits of man without which buildings and palaces are but gross handyworks . . .

FRANCIS BACON (1561–1626)
'Of Gardens'

To own a bit of ground, to scratch it with a hoe, to plant seeds and watch the renewal of life – this is the commonest delight of the race, the most satisfactory thing a man can do.

CHARLES DUDLEY WARNER
(1829-1900)
My Summer in a Garden

HOE GENTLY BECAUSE THE seeds from weeds – of which there can be hundreds – will germinate again if you dig up too much soil and bury them.

Hoeing: A manual method of severing roots from stems of newly planted flowers and vegetables.

HENRY BEARD (N.D.)
American humorist

THERE ARE PRAIRIE GRASSES in the Midwest whose roots stretch dozens of miles.

There are several ways to lay out a little garden; the best way is to get a gardener.

KAREL ČAPEK (1890–1938)
Czech writer and inventor of the word 'robot'

USING A POWER MOWER, WITH three or four horsepower, for one hour to cut the grass releases the pollution equivalent of driving 350 miles in a car.

IT IS BEST TO HOE IN DRY weather, when weeds are at their most vulnerable and easier to remove.

A substitute for mowing with the scythe has lately been introduced in the form of a mowing machine which requires far less skill and exertion than the scythe, and answers perfectly where the surface of the soil to be mowed is perfectly smooth and firm, the grass of even quality, and the machine only used in dry weather. It is particularly adapted for amateurs, affording an excellent exercise to the arms and every part of the body; but it is proper to observe that many gardeners are prejudiced against it.

JANE LOUDON
The Ladies' Companion to the Flower Garden, 1841

In one hour the average **sprinkler** uses the equivalent of two days' water consumption by a family of four. The average family uses approximately 500 litres of pure water per day.

Fresh water is a precious resource. Only 2.5 per cent of the world's **water** is not salty, and of that two-thirds is locked in icecaps and glaciers. Of the remaining amount, 20 per cent is in areas too remote for human access, and of the remaining 80 per cent about three-quarters comes at the wrong time and place – in the form of monsoons and floods – and is not captured for use by people. The remainder is less than 0.08 of 1 per cent of the total water on the planet.

World Water Commission, 2000

A garden is never so good as it will be next year.

Thomas Cooper (n.d.)

Edwin Budding, the English inventor of the lawnmower, said of his brainchild: *Country gentlemen may find in using the machine themselves, an amusing, useful and healthy exercise.*

When starting your garden seedlings indoors, plant the seeds in the halves of **eggshells**, which are a good natural fertilizer. Crack the shells around the roots of the young plants when you transplant them outdoors. If you want to revive an ailing houseplant, take some empty eggshells (an excellent source of calcium), dry them out in the oven and crush them into a powder. Add the powder to a small jug of water and then leave it for a day or so before pouring into the plant pots.

The UK water industry has 1,584 boreholes, 666 reservoirs and 602 river abstractions. Two-thirds of our water comes from surface water and one-third from groundwater.

The good rain, like the bad preacher, does not know when to leave off.

Ralph Waldo Emerson (1803–82)

TRY TO WATER YOUR PLANTS, especially the younger ones, with **warm water**, as cold water straight from a tap can give them an unpleasant shock. If you do not have water butts around your garden, or if they are empty after a prolonged dry spell, fill up as many watering cans and buckets as you can and leave them standing around to warm up, preferably in a greenhouse.

God made rainy days so gardeners could get the housework done.

ANON

I think the true gardener is the reverent servant of Nature, not her truculent, wife-beating master.

REGINALD FARRER
In a Yorkshire Garden, 1909

A Gardener's Work is never at an end; it begins with the year, and continues to the next.

JOHN EVELYN (1620–1706)

Gardening is an even more relentless and unending battle than the politics of Westminster.

FORMER CONSERVATIVE MINISTER
JOHN BIFFEN (B. 1903)
in *The Times*

Six ways to conserve water in your garden

1 Put plants with similar moisture requirements close together so that you can water them in one good dousing rather than having to visit various different areas to give them all a good soaking.

2 Mulch your plants with compost, straw, dead leaves or bark chips. You can also encircle the bases of larger plants with stones, which stops them from drying out quickly in hot weather.

3 Improve the water-retaining qualities of your earth by digging in lots of organic matter.

4 Place thirstier plants in positions where they will benefit from water running down any slopes you may have.

5 Place drought-tolerant plants like lavender and rosemary in positions where they will shelter their thirstier friends from the dehydrating effects of the wind.

6 Don't water your lawn.

No man feels but more of a man in the world if he have but a bit of ground that he can call his own. However small it is on the surface it is 4,000 miles deep and that is a very handsome property.

Charles Dudley Warner
(1829–1900)

The word **gazebo** is thought to be an abbreviation for 'gaze about'. Others claim it is a bastardization of Latin, **gazebo** meaning 'I shall gaze'.

In March and in April, from morning to night:
In sowing and setting, good housewives delight
To have in their garden or some other plot . . .
To trim up their house, and to furnish their pot.

At midsummer, down with thy brambles and brakes:
And after abrode, with thy forks and thy rakes.
Set mowers a work, while the meadows be grown
The longer they stand, so much more to be mown.

Thomas Tusser
A Hundred Good Points of Husbandry, 1557

But each spring . . . a gardening instinct, sure as the sap rising in the trees, stirs within us. We look about and decide to tame another little bit of ground.

Lewis Gantt (n.d.)

The man who has planted a garden feels that he has done something for the good of the world.

Vita Sackville-West
(1892–1962)

We have heard it said that the labourer who toils in his garden, expends the strength which should be husbanded for his master's service: to which cold-blooded dictate of a heart worthy of a slave-owner we reply that all experience demonstrates that those labourers, who devote their leisure to their own gardens, are invariably the best characters and the best workmen in a parish.

GEORGE W. JOHNSON
The Gardeners Almanack, 1844

WINDOW BOXES HAVE MUCH TO recommend them: they look great and give your house a bit of character, breaking up the symmetry of straight lines and softening the view from outside. They also smell good if you grow some strongly scented plants that produce wafts of pleasant fragrances through an open window on a summer's day. The smell of certain common herbs will also make flies and mosquitoes think twice before entering the house. If you live in a flat or a townhouse with a small garden, and you miss the natural experience enjoyed by country dwellers, window boxes have the added attraction of encouraging wildlife (albeit on a small scale), such as birds and bumblebees. The yellow and white 'poached egg flower', for instance, will attract hoverflies and lacewings, which eat aphids. If you want a window box for your kitchen that is both useful and attractive, there are dozens of herbs you can consider growing including parsley, oregano, thyme, chives, rosemary, marjoram, basil and sage. The one drawback with growing anything in containers is that the soil dries out quicker than if it was in an open site, and you need to water them fairly often. One way round this is to grow plants that enjoy dry conditions, such as lavender, thyme and rosemary. If not, make sure you have a layer on top of moisture-retaining material like compost, manure and/or bark chips. (For more tips about retaining water, see the entry about hanging baskets on p. 28.)

You may detest the **weeds** in your garden with a burning passion, but you have to take your hat off and salute their **incredible guile** and resourcefulness. Weeds didn't get where they are today by letting the fancier plants in the neighbourhood walk all over them. On the contrary, over the millennia, and without the slightest help or encouragement from man, they have evolved and adapted in a number of ways to ensure that they can more than hold their own with their more sophisticated cousins. For a start, they tend to grow much faster than other plants, often strangling or suffocating any rivals in the vicinity, or at the very least stealing their light, nutrients and moisture. They also make themselves unattractive to many insects and animals. Then they spawn an enormous family in the form of hundreds, even thousands of seeds, which they quickly send out into the wide world in a host of ingenious ways. The seeds are often tough and durable and can lie dormant for years.

March isn't the only thing that's in like a lion and out like a lamb.

Mae West (1893–1980)

10 Capability Brown Gardens

Blenheim, Oxfordshire
Broadlands, Hampshire
Chatsworth, Derbyshire
Ickworth, Suffolk
Longleat, Wiltshire
Luton Hoo, Hertfordshire
Petworth House, West Sussex
Sheffield Park, East Sussex
Syon Park, London
Warwick Castle, Warwickshire

A house though otherwise beautiful, yet if it hath no Garden belonging to it, is more like a Prison than a House.

William Coles, 1656

The pleasure gardening in England is the article in which it surpasses all the earth.

Thomas Jefferson 1786

Tomorrow is the busiest day of the year.

SPANISH PROVERB

NEW BRICKS, WALLS, ROCKS or stonework in the garden often look unsightly amongst older-looking, more weathered materials. You can hasten the ageing process by smearing the new additions with **live yoghurt**, the active bacteria of which quickly attract moss and lichen as well as darkening the surface of the stone.

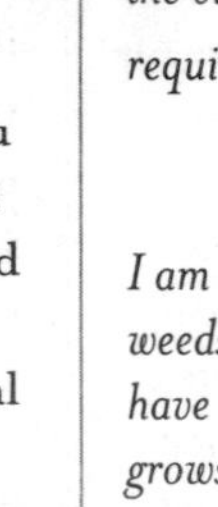

LATE FROSTS CAN STRIKE EVERY part of Britain, and they often occur in the areas furthest from the sea (the Midlands) and on the higher ground. If the Met Office is forecasting a late frost there are a number of provisions you can make to protect vulnerable plants. Old newspapers, pegged into the ground or weighted with stones, straw, horticultural fleece or any other strips of natural material that can 'breathe' should do the trick.

Half the interest of the garden is the exercise of the imagination. You are always living three, or indeed six, months hence. I believe that people entirely devoid of imagination never can be really good gardeners. To be content with the present, and not striving about the future, is fatal.

C. W. EARLE ('MRS EARLE')
Pot Pourri from a Surrey Garden, 1897

YOU SHOULD FEED THE SOIL around a plant so that its roots get to absorb the goodness. Don't feed at the foot of its stem, as this encourages **'soft' growth** and makes it more vulnerable to diseases and pests.

The [English] *climate is unpleasant, with frequent rain and mist, but it does not suffer from extreme cold. The soil is fertile and is suitable for all crops except the vine, the olive and other plants requiring warmer climes.*

TACITUS (55–120)

I am told that abundant and rank weeds are signs of a rich soil, but I have noticed that a thin poor soil grows little but weeds.

CHARLES DUDLEY WARNER
My Summer in a Garden, 1871

Trying to stuff piles of **autumn leaves** into a refuse sack can be a frustrating business, as you attempt to keep the bag open while using the other hand to scoop up the leaves. You can avoid blowing your top, and save yourself some time by freeing up both hands, if you take an old cardboard box, cut out its bottom and insert it into the bag to create a wide-open top. It's best to gather the leaves as soon as possible before they get wet and harm your lawn. Tie up the bags and make a bowl-shaped area at the top to collect rainwater, then cut a few small holes in the plastic so that the water can seep in. Make a few more around the bottom and sides to allow access to beneficial insects. Leave for 12 months, after which you will have a wonderful rich mulch. The result is a clear garden and a valuable source of nutrients for your plants the following year.

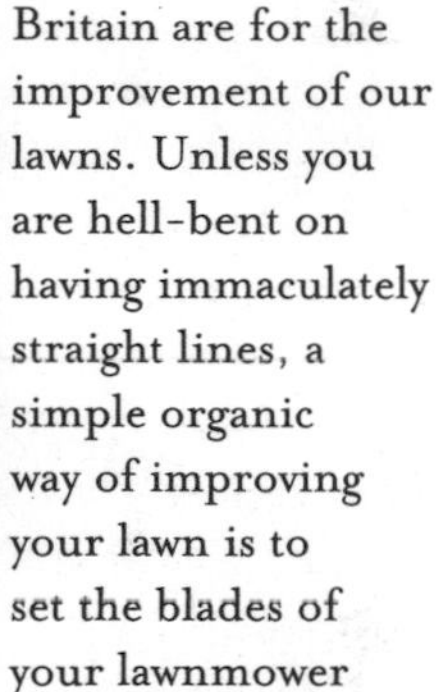

On early spring: *The vegetable world begins to move and swell and the saps to rise, till in the completest silence of lone gardens and trackless plantations, where everything seems helpless and still after the bond and slavery of frost, there are bustlings, strainings, united thrusts, and pulls-all-together, in comparison with which the powerful tugs of cranes and pulleys in a noisy city are but pigmy efforts.*

Thomas Hardy
Far from the Madding Crowd, 1874

There is no ancient gentlemen but gardeners.

William Shakespeare (1564–1616)
Hamlet

Over three-quarters of all **garden chemicals** sold in Britain are for the improvement of our lawns. Unless you are hell-bent on having immaculately straight lines, a simple organic way of improving your lawn is to set the blades of your lawnmower a little higher than you would do ordinarily and leave off the collecting box. The cut grass is packed with goodness when it rots down, and the teeming insect life in the lawn will quickly set to work to pull all the clippings down to the ground and out of view.

One of the most delightful things about a garden is the anticipation it provides.

W. E. Johns (1893–1968)
The Passing Show

SPRING

April comes like an idiot, babbling and strewing flowers.

EDNA ST VINCENT MILLAY (1892–1950)

Spring vegetables

Asparagus, lettuce, nettles, purple sprouting broccoli, radishes, rhubarb, rocket, sorrel, spinach, spring greens, spring onions

He that is in a town in May loseth his spring.

GEORGE HERBERT (1593–1633)

Every year back spring comes, with nasty little birds, yapping their fool heads off and the ground all mucked up with plants.

DOROTHY PARKER (1893–1967)

Spring travels in a north-easterly direction across the UK, taking roughly eight weeks to spread upcountry from Cornwall to the northern coast of Scotland.

In the spring, at the end of the day, you should smell like dirt.

MARGARET ATWOOD (B. 1939)

Spring is sooner recognized by plants than by men.

CHINESE PROVERB

Spring Birthdays

9 March 1763: William Cobbett, Farnham, Surrey
9 March 1892: Vita Sackville-West, Knole Castle, Kent
12 March 1613: André Lenôtre, Paris
8 April 1783: J. C. Loudon, Cambusland, Scotland
2 May 1949: Alan Titchmarsh, Ilkley, Yorkshire

The first day of spring was once the time for taking the young virgins into the fields, there in dalliance to set an example in fertility for nature to follow. Now we just set the clocks an hour ahead and change the oil in the crankcase.

E. B. White (1899–1985)

The March wind roars
Like a lion in the sky,
And makes us shiver
As he passes by.

When winds are soft,
And the days are warm and clear,
Just like a gentle lamb,
Then spring is here.

Anon

Spring – an experience in immortality.

Henry D. Thoreau (1817–62)

A little Madness in the Spring
Is wholesome even for the King.

Emily Dickinson (1830–86)

So many seeds, so little time.

Anon

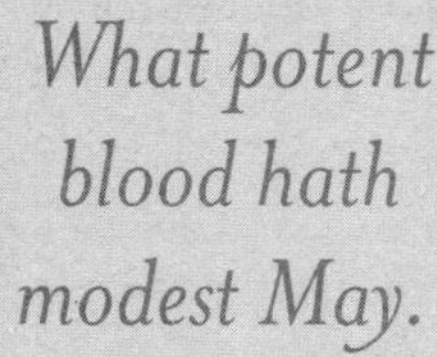

What potent blood hath modest May.

Ralph W. Emerson (1803–82)

The five worst things about spring gardening

1. Slug invasion
2. Packed garden centres
3. Late frosts
4. Back-breaking work for a month or so
5. Realizing in early summer that you have forgotten to sow or set certain plants

April is the cruellest month, breeding
Lilacs out of the dead land, mixing
Memory and desire, stirring
Dull roots with spring rain.

T. S. Eliot
'The Waste Land', 1922

Vegetables

Sex is good but not as good as fresh, sweet corn

CHEWING **PARSLEY** WILL HELP get rid of **bad breath** as it reduces the production of intestinal gas by improving digestion. Parsley is very high in chlorophyll, and though it smells strong, its taste is mild. Chewed after a meal, it is particularly effective in neutralizing the foul odours of onions and garlic. Chewing mint, basil, rosemary or thyme achieves a similar result.

ONIONS HAVE BEEN USED TO treat **gunshot wounds** since the sixteenth century. During the American Civil War, General Grant refused to move his Union troops without supplies of onions.

ONION AND GARLIC BULBS are best hung on string or in nets in a cool airy place indoors where they will not be susceptible to frosts. (You can also use old **tights and stockings.**) Stringing is fairly straightforward: after they have been allowed to dry out, remove the loose skin, roots and most of the tops. Tie four or five onions together by their stalks, and then fasten the whole bunch to a longer piece of string, but not so long that it will snap if you make the load too heavy. Hang the string from the ceiling or a hook rack, and then add more onions by tying them on one by one and sliding them down to the bunch at the bottom.

Do not eat garlic or onions, for their smell will reveal that you are a peasant.

CERVANTES (1547–1616)
Don Quixote

IF YOU ARE PLANNING TO FREEZE some excess **broccoli** from the garden, first place the pieces on a tray and put them in the

Britain's Favourite Vegetable

1 Onion
2 Sweetcorn
3 Carrots
4 Pepper
5 Brussels sprouts
6 Garlic
7 Peas
8 Spinach
9 Beans
10 Courgette

(According to a nationwide poll conducted in 2005)

freezer, otherwise they will form a solid lump. When they are frozen, you can safely bag them or put them in a container.

FOR A QUICK, DELICIOUS **homemade coleslaw**, shred half a large cabbage, a carrot and an onion, stir in some mayonnaise, two tablespoons of soft brown sugar, a sprinkling of salt and pepper and either half a wine glass of vinegar or a teaspoon of mustard powder. Refrigerate before eating.

The juice of onions anointed upon a bild or bald head in the Sun, bringeth the haire again very speedily.

JOHN GERARD (1545–1612)

ONION SOUP, FRENCH-STYLE

Gently fry about eight medium-sized chopped onions in 50g melted butter in a large frying pan, add a big tablespoon of soft brown sugar and stir until the onions are soft, light brown and slightly caramelized ❁ Pour the onions into a large saucepan, and add a glass or two of white wine and 1 litre of stock, preferably beef, and let it all simmer for about 15–20 minutes ❁ You can add some large croutons as they do in France, but if you don't enjoy great lumps of flaccid sponge floating about in your soup, just add some salt and pepper and sprinkle on some grated cheese.

AMERICANS CONSUME 30 pounds of **potatoes** per person per year, 25 per cent of which is in the form of French fries.

WHEN YOU **CHIT A POTATO**, all you are doing is stealing a march on nature by encouraging early growth. Before planting out, rub off all but the two strongest shoots. Tradition has it that we plant our potatoes on Good Friday, but as that is a movable feast day it is not advisable to observe this convention religiously. When it comes to second-guessing the last frosts, it all depends on your local conditions, such as how far north, how high up and how far from the sea you live. If ever in any doubt about when to plant or sow, the best bit of advice you will get is from a gardener who has lived in your area for a long time. If your potatoes are already up when a **late frost** is forecast, cover them with newspapers or some kind of fleece.

THE WORD **AUBERGINE** DERIVES from the Sanskrit word *vatin-ganah*, meaning 'go away wind' or **anti-flatulence** vegetable. It is said that aubergine is excellent for treating intestinal gas.

SPINACH CONTAINS NO MORE IRON than most other green vegetables. Its reputation for imparting exceptional Popeye-style strength arose at the end of the nineteenth century when a leading scientist was researching its nutritional properties. When noting its iron content, he put the decimal point in the wrong place.

French beans and runner beans need plenty of water when they come into flower. To prevent them from drying out, perhaps if you are away from home for a few days, add some mulch, or even some old carpet cut to size, around the base of the plants. It is very likely that at some point in the summer you will have more green beans than you can shake a stick at. If you are bored with eating them boiled or steamed, here are three very simple recipes to help you enjoy the glut. As with so many easy but delicious recipes, you don't have to be overly fussy about the precise amount of ingredients.

FRENCH BEAN SALAD

You can make a great big salad out of your glut of beans with the following, idiotically easy dish, which works particularly well as an accompaniment to a meat-heavy barbecue ✿ Top and tail as many green beans as you need, boil them until tender, plunge them into cold water and transfer them to a serving bowl ✿ Pick a variety of herbs from the garden, chop them all up in a mug using scissors and sprinkle the mixture over the beans ✿ At the last minute, pour over a basic vinaigrette ✿ If you're feeling adventurous, you can add a few anchovies or some chopped fresh green chilli to give it a kick.

GREEN BEANS PROVENÇAL-STYLE

Heat some olive oil, shallots and garlic in a large casserole-type pan for a few minutes before adding some peeled and chopped tomatoes (or a tin of ready-prepared ones if you are in a hurry) and then heat until most of the liquid has evaporated ✿ At the same time boil or steam your beans, which are best chopped in half, before adding them to the tomato mix.

FRENCH BEANS AND WILTED LETTUCE

For a surfeit of lettuce, the French have a delicious traditional recipe that combines it with the beans ✿ Boil and drain your beans, while at the same time frying some garlic and thinly sliced onions until they are soft and then adding some shredded lettuce ✿ After a minute or so, when the lettuce has wilted, add the beans.

THERE ARE MANY DIFFERENT COMBINATIONS OF PLANTS THAT WHEN grown in close proximity to each other are known or thought to be mutually beneficial. They attract good insects or deter pests, offer shade, loosen the soil for the other's roots or simply enjoy the same kind of growing conditions. Marigolds, with their great capacity to deter all manner of pests, are perhaps the best-known friend of the vegetable. They are particularly welcome near potatoes, tomatoes, asparagus and cabbage. The following is a selection of good combinations but it is not a complete list:

Happy bedfellows

Artichoke – parsley

Asparagus – tomatoes, parsley, basil

Beetroot – onion

Dwarf beans – potatoes, sweetcorn, celery, cucumber

Broad beans – cabbage, carrots, potatoes, cauliflower

Runner beans – sweetcorn

Broccoli – potatoes, thyme, sage, mint, rosemary, dill

Cabbage – potatoes, thyme, sage, mint, rosemary, dill

Carrots – onions, garlic, leeks, broad beans, lettuce, tomatoes

Cauliflower – broad beans, thyme, sage, mint, rosemary, dill, potatoes

Courgette – borage, nasturtium

Cucumber – borage, lettuce, sunflower, peas

Garlic – carrots, beetroot, strawberries, tomatoes, lettuce, raspberries

Lettuce – carrots, cucumber, garlic, strawberries

Marrow – sweetcorn, nasturtium, fennel

Peas – sweetcorn, turnips, mint, radish, all three bean types

Potatoes – cabbage, mint, parsley, lavender*

Pumpkin – sweetcorn, nasturtium

Radish – cucumber, peas, lettuce, nasturtium

Spinach – peas, strawberries, rhubarb

Tomatoes – asparagus, carrots, celery, onion, parsley

Turnips – peas, radish

**Horseradish is also very good, but plant it in pots nearby or it will rampage through your vegetable plot and you'll never get rid of it*

10 combinations of plants that don't like each other

Artichoke/garlic
Beetroot/runner beans
Broccoli/strawberries and tomatoes
Cabbages/strawberries
Cabbages/tomatoes
Cauliflowers/tomatoes and spinach
Cucumber/potatoes
Garlic/peas and beans
Lettuce/fennel
Onions/all beans
Peas/potatoes
Potatoes/tomatoes
Potatoes/pumpkins and marrow
Radish/potatoes
Tomatoes/fennel

THE GREAT ROMAN EMPEROR **Tiberius** (ruled AD 14–37) was an avid eater of cucumbers, which were reputedly served to him with every meal. To meet the imperial demand, his gardeners became experts at forcing them in wheeled beds which could be moved around into the sun.

SOME SAY THAT THE STRONG odour of the camphor from crushed-up mothballs deters **carrot fly** more effectively than onions. Planting a **mothball** next to each tuber is thought to protect pototoes from attack. Another good way to protect carrots from carrot fly is to scatter a few creosoted rags around them (this protects cabbages too). Creosote also deters deer and badgers, both of them rapacious destroyers of gardens, but the smell is also unpleasant to most humans.

Cabbage root fly will happily destroy your crop if you let it, but you can deter these pests without resort to beastly chemicals by adding some chopped-up rhubarb leaves to the hole before planting. If a cabbage does fall victim to the root fly, leave it in the ground, otherwise the pests will simply move on to the next plant.

IF YOU HAVE ANY ANXIETIES ABOUT whether your **seeds** are too old or too poor to germinate properly, sprinkle a few of them on a damp piece of **kitchen towel** and keep in a warm place to see if they germinate.

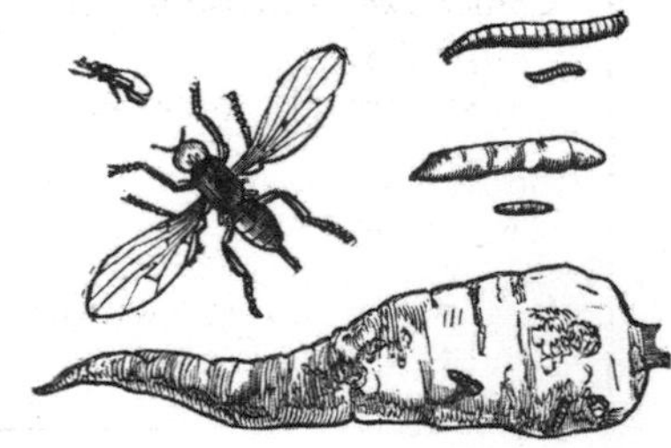

THE FOLLOWING 10 EXCERPTS are from Ministry of Agriculture booklets published as guides for amateur gardeners towards the end of the Second World War, when rationing was likely to remain in place for some time. The advice holds good today.

Once weather and soil are right, we should take time by the forelock and get on with the job – not leaving everything to the weekend if we can help it, but seizing any opportunity of an evening when it's fine to put in a little time on essential work on the plot.

Unfortunately, despite many warnings, some amateurs have been taken in every year by unscrupulous people who sell them tomato plants far too early for planting outside. It is foolish to hope that the danger of frost is past until at least the end of May.

If you suffer badly from wireworm, it is worth trying to trap them. An old potato makes a good trap, or three inches of old kale or Brussels stalk split down the middle. Put these traps a few inches below ground in spring, marking the spots with sticks. You can do a great deal to rid yourself of wireworm if you set traps regularly.

There are some people who seem to think that the compost heap is a new idea, introduced because farmyard manure is hard to come by. It is no novelty, for the gardening books of a century or more ago mentioned it; long before it was called 'compost' the value of decayed vegetable refuse was well known and understood, particularly by the professional gardener.

Now the black fly's bitterest foe is the ladybird, but although she makes all her meals off black or green flies, she cannot cope with all of them. The black fly usually attacks the top of the plant first, just when it is beginning to flower, so pinch off the top to check it. The ladybird won't mind.

Early-planted Brussels sprouts should now [October] *be ready for picking. There is a right way and a wrong way of gathering them. Start at the bottom and clear the stem of sprouts as they become large enough; don't pick a sprout here and there, but do it systematically from the bottom of the stem.*

At least 10lb of tomatoes are required to produce 1oz of seed. Remove from the fruit the pulp containing the seeds and put it in a jar to ferment. After two or three days, tip it into a fine sieve and wash it vigorously under the tap; the pulp will wash away from the seeds, which may then be spread on muslin to dry.

Don't kill the centipede, for it goes for your enemies – small slugs, worms and insects. The friendly centipede moves very quickly, while the millipede – a nasty sort of chap – moves slowly, though he has got two pairs of legs to every section, as against the centipede's one. You cannot go far wrong if you kill the slow movers and let the fast movers live. Anyhow, it's death to the millipede that attacks the roots of most of your plants!

Marrows and pumpkins: These may be stored for winter use as vegetables and for preserving. Only fully developed and ripened fruits should be set aside for storage, and they should be handled carefully to avoid bruising the skins. Being very susceptible to low temperatures and easily damaged by frost, these fruits need a warm, dry atmosphere, such as that of a kitchen, bedroom or attic, to ensure successful storage . . . The fruits may be placed in crates or boxes or laid out singly on shelves, but they are best hung from the ceiling in nets. Given this treatment, they can usually be relied upon to keep in good condition until January or February.

Some people say, 'Marry in May, repent alway.' Perhaps if we do marry in May we may find the maid, like the month, fickle and fitful; sometimes sunny, sometimes stormy and sometimes more than a bit frosty! That is the trouble with May, those killing frosts that do so much damage to our fruit blossom and young potato plants, and catch the unwise and unwary who put out their tomato plants too early and without protection.

Tomatoes and squash never fail to reach maturity. You can spray them with acid, beat them with sticks and burn them; they love it.

S. J. Perelman (1904–79)

IT IS A GOOD IDEA TO **BLANCH your vegetables** in boiling water, but only for a minute or two at the most, before you freeze them. This stops the enzymes getting to work on decomposing your produce. Make sure the vegetables have cooled completely before freezing.

Am I to be sacrificed, broiled, roasted, for the sake of the increased vigour of a few vegetables? The thing is perfectly absurd. If I were rich I think I would have my garden covered with an awning, so that it would be comfortable to work in. Another very good way to do, and probably not so expensive as the awning, would be to have four persons carry a sort of canopy over as you hoed. And there might be a person at each end of the row with some cool refreshing drink.

CHARLES DUDLEY WARNER (1829–1900), *My Summer in a Garden*

EATING **CELERY** RESULTS IN negative calories: it takes more **calories** to eat a piece of celery than the celery has in it to begin with.

THIRSTY VEGETABLES SUCH AS runner beans and celery like cool soil and moist ground, so it's a good idea to plant them in a trench with water-retaining material such as old newspapers to help them through long dry periods.

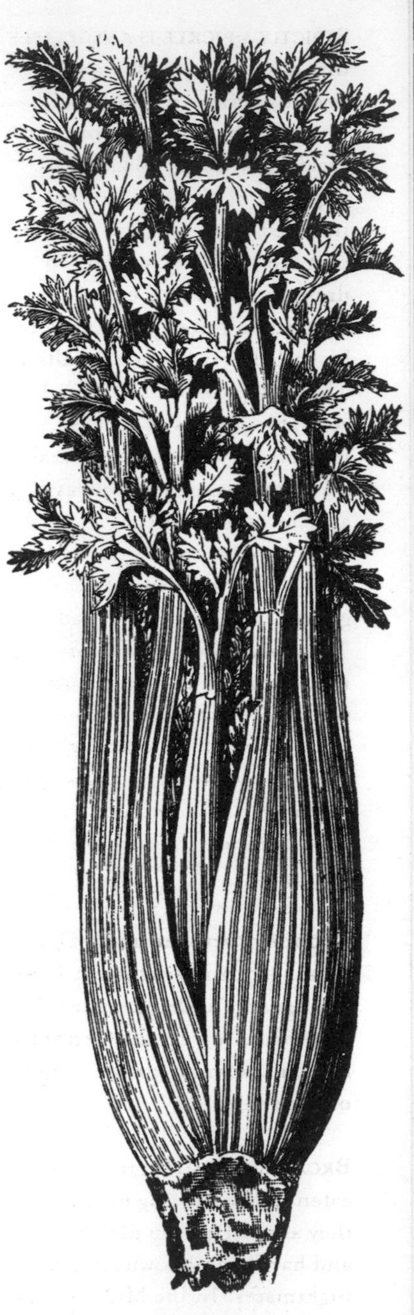

Strictly a **pickle** is a vegetable that has been kept in vinegar, as distinct from a **chutney**, which has been cooked in it. Pickling is simple. Soak your vegetable of choice overnight in salted water (roughly one part salt to 10 parts water) to draw out some of the moisture. (Beetroots and peppers need to be boiled first for best results.) Then simply add the drained vegetables to jars of **vinegar** made from wine, cider or malt. To boost flavour, add spices and herbs of your choice. The most popular are mustard seeds, black peppercorns, garlic cloves and celery seeds. There are thousands of easy recipes on the Internet, but you may want to experiment for yourself. It is good kitchen practice to sterilize the jars by placing them mouth up in a pan of simmering water for 15 minutes.

Peppers contain up to five times as much vitamin C as oranges. The highest levels are found when the peppers are in the early green stage of ripening.

Broad beans should not be eaten in the evening because they are difficult to digest, and have been known to cause nightmares. In the Middle Ages it was thought that the souls of the dead lived inside the humble broad bean because it disturbed people's sleep. The Greek philosopher and mathematician **Pythagoras** would not eat broad beans under any circumstances, so convinced was he that they were possessed by evil spirits.

To get the best results you must talk to your vegetables.

Prince Charles
Observer, 1986

There are more **greenhouses per capita** in Britain than in any other country in the world.

The alkaloid capsaicin is the ingredient that gives **peppers** their heat, and its quantity is much greater in the hot peppers than the bell variety. Climate conditions and geographic location are also factors in the intensity of the heat. In short, the hotter the weather, the hotter the pepper. Peppers left to reach maturity have a higher capsaicin content than those picked early.

An old ironing board makes an excellent potting table. You can adjust it to the required height and it takes up next to no room when you store it away.

FOR MUCH OF THE NINETEENTH century, the '**baked potato man**' was a common sight on London streets, selling hot spuds in their jackets to passers-by. It is estimated that at the height of their popularity there were over 250 itinerant potato men plying their trade. Most English gentlemen wouldn't be seen dead eating a vulgar potato off the street, but they often bought them as hand-warmers during the harsher winters of that time.

WILLIAM COBBETT, THE nineteenth-century writer, radical and agriculturist and a man of forthright opinions, was not a great admirer of the potato. *Leave Ireland to her lazy root . . . It is the root also of slovenliness, filth, misery and slavery; its cultivation has increased in England with the increase of the paupers: both, I thank God, are on the decline.*

FOR MANY YEARS, CENTURIES even, the **potato** was considered unsuitable for general consumption in Europe. The Spanish, who first came across the vegetable in South America in the middle of the sixteenth century, saw it as fit only for their slaves. The English later introduced potatoes as a staple food for the Irish, while preferring to eat parsnips with their roast beef back home. The Germans used it to feed animals and prisoners. Part of the potato's image problem lay in the fact that, like the tomato, it is a member of the nightshade family and its leaves bear a resemblance to those of its deadly cousin. Some even thought the potato caused **leprosy** on account of its lumpy, pock-marked appearance. Protestants, moreover, associated it with Catholicism because of its South American origins. The connection was so strong in the minds of many that in the mid-eighteenth century there was an election banner in Lewes, Sussex, which read, 'No Potatoes, No Popery'.

HOW THE **POTATO** REACHED Irish shores remains open to conjecture, but the popular belief that **Walter Raleigh** introduced it may well be true. What is certain is that the Irish welcomed it as a gift from God at a time when English armies were laying waste to their more visible crops and massacring communities in their efforts to subdue the country.

To raise potatoes for the purpose of being used instead of bread is a thing mischievous to the nation.

WILLIAM COBBETT (1763–1835)

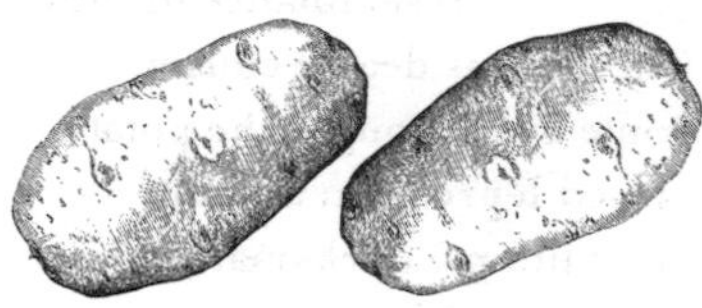

BASEBALL LEGEND **BABE RUTH** used to wear a **cabbage** leaf under his cap to keep his head cool in games.

POTATOES WERE BRITONS' principal source of vitamin C in the two world wars. Cases of **scurvy** soared in 1916 when part of the potato crop failed. Today, 90 per cent of the world's potatoes are grown in Europe.

FOLLOWING THE DEVASTATING **Black Death** of the mid-fourteenth century, which wiped out about a third of Europe's population, there were insufficient peasants in England to work the landowners' fields and gardens. To avert a food crisis, landlords leased small plots of land to tenants so that they might grow their own produce. It was thus, historians claim, that the famous English **cottage garden** came into existence.

IN THE NINETEENTH CENTURY German scientist Justus Von Liebig and Cornish scientist Sir Humphry Davy proved that phosphates and nitrates

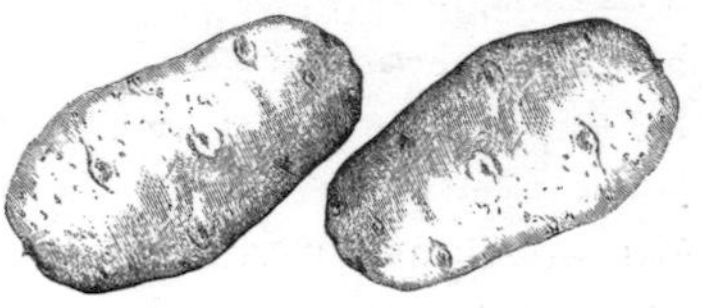

improved the soil, triggering a rush among farmers and gardeners for **old bones** that could be crushed into a powder and used as a fertilizer. Liebig accused British merchants of digging up the battlefields of Europe in their frantic search for bones, while tens of thousands of animal bones were imported from Eastern Europe to meet demand.

PINCH OUT THE TOPS OF **runner beans** when they have reached the top of the supports they have grown up, otherwise the plant will become top heavy.

AN ATTRACTIVE AND INGENIOUS way of planting **runner beans** is to grow them up sunflower stems. The sunflower seeds should be planted as early as possible so that they are strong enough to support the beans. Each **sunflower** can host up to three runner beans, but it is best probably to stick with two.

BROAD BEAN AND MINT PURÉE

Boil the broad beans for about five minutes and put in a food processor with some crushed garlic, lemon juice, a cup of chopped fresh mint, a few tablespoons of olive oil, salt and pepper ❁ You can serve the purée as a dip, but it also works well as an accompaniment for fish, and is particularly delicious as a 'bed' for scallops.

BLACKFLY APHIDS CAN CAUSE havoc on broad beans, but there are some good natural deterrents. Boil up two or three pounds of elder leaves in about two gallons of water and spray the liquid on once it has cooled. Planting marigolds nearby or between the rows also works.

Grow your own walking stick

For decades the **Giant Jersey cabbage** was cultivated in the Channel Islands for the manufacture of walking sticks, which were sold to Victorian gentlemen. The leaves of this tall, thick-stemmed plant were used in soups (and also fed to cattle), while the roots were often carved into thimbles. The production of the Giant Jersey has declined dramatically in last 50 years or so, but if you can get hold of some seeds they may be fun to grow as a novelty one year. When the plants are about 18 inches high, take off the lower leaves and the stalks should double in size by the end of their first season. In their second they will flower and at the end of their third they will reach heights of over **six feet**. Cut the stalks into walking stick lengths and dry them for six months. Varnish them with a protective coating and fit a handle to the top and a rubber stop to the bottom. Give the finished article to your grandfather for Christmas and he will realize that a) you are a creative genius and b) you are completely skint, can't afford proper presents from the shops and wouldn't mind a handout before he leaves.

THE WORD **AVOCADO** IS A corruption of a Central American word meaning 'testicles'. The fruit was thought to resemble a **scrotum**. Avocados (strictly a tropical fruit but used like a vegetable) were extremely rare in Britain until after the Second World War, when Israel began to produce them on a huge scale.

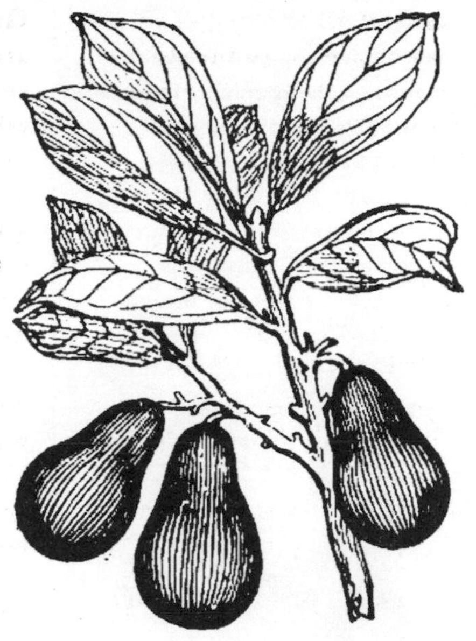

THE EMINENT SIXTEENTH-century Swiss doctor **Paracelsus** believed the healing properties of plants were carried in their natural signature: that, for example, rock-breaking roots such as saxifrage cured kidney stones and the ear-like leaves of cyclamen cured earache.

To make gherkins, scrub your small cucumbers with a hard-bristled brush without breaking the skin to get rid of the sharp prickles, then immerse them for 24 hours in well-salted water. Drain and wash them and pack into jars filled with the pickling vinegar of your choice. Add spices according to taste if you wish and then seal tightly. Like most pickled vegetables, they taste better if left for a couple of months.

AMERICAN COLONISTS FROM the *Mayflower* baked their pumpkins whole in the ashes of a fire. Once the **pumpkins** were cooked, they cut them open and served them with animal fat and maple syrup. If the thought of this doesn't get you drooling, here is a very simple recipe for pumpkin pie. Line an ovenproof dish with some shortcrust pastry, cover it with slices of pumpkin and sprinkle over them two or three handfuls of currants and some sugar with a teaspoon or so of cinnamon. Cover the dish with more pastry and cook it in a hot oven for about half an hour. You can serve it hot with cream or eat it cold.

YOU COULD BE DOING AS much harm as good to the environment by buying **organic vegetables** because most of those in British shops have been imported, many of them from thousands of miles away. By growing your own you will be doing an extra favour to yourself and your family because freshly picked produce contains far higher amounts of nutrients, vitamins and antioxidants.

TO KEEP **BIRDS AND MICE** FROM your freshly sown seeds, rub them with a cloth moistened in **paraffin**. Birds and mice hate the smell and taste. This is especially effective for broad beans, peas and mangetout. It is also a good idea to cover newly sown seeds with sprigs of gorse or thorns. Failing that, staking down some finely meshed chicken wire also works.

You can bury a lot of troubles digging in the dirt.

ANON

SOWING SMALL SEEDS SUCH AS carrots in an orderly way can be troublesome. For a more even sowing, cut a small hole in the corner of an envelope and then tap it gently as you work your way along the drill. Or use a small thin plastic pot with a few little holes punched in the bottom to enable you to sprinkle the seed evenly and thinly over the ground as you walk along.

THERE ARE DOZENS OF homespun tips for **deterring cats** from using your plot as a giant lavatory and one of the most popular is sprinkling some kind of hot pepper solution around your vegetables. This is all very well, but if it rains you will have to do it all over again, which could get boring for those of us living in Britain and Ireland. Also, if the solution is too powerful you can cause great distress to cats if they get the pepper in their eyes when washing their paws. There are even stories (possibly apocryphal) that cats have experienced such agony that they have scratched their own eyes out. One safe way to deter cats from fouling and digging up beds is to fill up old bottles with a **vile-smelling liquid** like Jeyes fluid or bleach and bury them up to their necks. Others swear by scattering around the peel of citrus fruits, which cats are said to hate. Keeping a spray bottle to hand near your plot and squirting the intruders with water whenever they saunter over is another effective means of control. Socks will slowly get the message that there is only one place for him to carry out his post-prandial poo, and that's next door in Mr Prendergast's garden.

PEA AND MINT SOUP

John Major's *Spitting Image* caricature may never have tired of eating plain boiled peas, but for a change he would have done well to persuade his wife 'Norman' to make a quick and delicious soup with them ❁ Boil about half a kilo of freshly podded peas in about a litre of vegetable or chicken stock along with a largish potato and an onion, both chopped roughly ❁ When the ingredients have softened, whiz them in a blender and then return the mixture to the pan and stir in some cream and a small handful of chopped fresh mint.

WITH **LARGER SEEDS** LIKE THOSE of the broad bean or the pea, there is a simple, age-old method for determining their quality: fill a tray with water and put in the seeds. The best seeds will sink, the inferior ones will float.

GERMINATION AND GROWTH will be enhanced if newly sown seeds are immediately watered with a can of very warm water, using a watering can with **a fine sprinkle**.

BORIS **PASTERNAK** WON the Nobel Prize for Literature for *Dr Zhivago*. Students of literature may be interested to know that *pasternak* is the Russian for **parsnip**. Meanwhile followers of European politics may wish to learn that *kohl* (as in former Chancellor Kohl) is German for cabbage.

EASY HOMEMADE PICCALILLI

Take 2.7kg vegetables, 500g salt, 220g sugar, 25g dry mustard, 25g ground ginger, 15g turmeric, 1.5 litre vinegar, 30g cornflour (the spice element is up to you) ❁ Cauliflower is the main ingredient of this very English relish but you can include an assortment of other vegetables such as cucumber, green tomatoes, runner beans and small onions ❁ Soak them in heavily salted water for 24 hours ❁ Wash, drain and cut them into sizes your mouth can manage ❁ Put the sugar, spices and vinegar into a large pan and heat until the sugar has dissolved ❁ Add the vegetables and simmer for a short while, making sure that they remain crispy ❁ Mix the flour with a small amount of vinegar and add to the pot ❁ Boil for about 2–3 minutes to cook the flour before transferring to jars ❁ Best left for two months or so.

BRITAIN'S RICHARD HOPE HOLDS the world record for growing the **longest beetroot**, stunning visitors to the 2003 Llanharry Giant Vegetable Championships in Wales with a whopper measuring 6.146 metres.

JERUSALEM ARTICHOKES have nothing to do with Jerusalem and nothing to do with artichokes. They were introduced into Europe from Canada by Samuel de Champlain in the seventeenth century and originally called Canadian artichokes (Samuel said they tasted a bit like artichokes and the word stuck, apparently). The origin of the name Jerusalem is an English corruption of the Italian *girasole*, meaning 'sunflower', which it resembles when in bloom. If you are wondering why it is that you tend to break wind after eating a Jerusalem artichoke, it is because its carbohydrates come mostly in the form of inulin, a sugar that can cause flatulence.

FRESH TOMATO SAUCE

You can use this sauce in any number of ways: as a simple pasta sauce, or for serving with baked chicken breasts or white fish ❁ Heat a knob of butter in some olive oil before adding a few teaspoons of crushed garlic, some chopped onion (and celery if you have some) ❁ After a few minutes pour in about a kilo of chopped and peeled tomatoes (with some herbs for added flavour if you wish), and heat for about a quarter of an hour ❁ If you have too much for immediate use, freeze the remainder as a whole or in separate bags of different quantities, depending on the size of your household.

Green tomatoes are delicious fried, but they will ripen if wrapped in tissue paper or muslin and left in a box, cupboard or drawer. The ripening process will be enhanced if you add a fully ripe one because it emits a hormone called ethylene that stimulates the others. Banana skins perform a similar ripening function, which will explain the curious sight of the yellowy black skins hanging over tomato plants in the greenhouses of people you may have previously regarded as insane.

Green tomato chutney is one of life's great pleasures. Take two kilos of chopped tomatoes, half a kilo of apples (peeled, cored and chopped), 250g of raisins, some chopped-up chillies and ginger, roughly half a kilo of chopped shallots, half a litre of pickling vinegar, half of a kilo of soft brown sugar, a couple of teaspoons of salt – and then simply pour the whole lot into a large pan and simmer until you have achieved the consistency you want. There is really only one thing to do with green tomato chutney and that is to eat it with some fresh bread and good hard cheese, ham or tongue, but if you want to surprise your neighbours you could always smear it over your naked body and run round the garden singing the Marseillaise.

Five reasons not to dig your vegetable plot

1 Digging encourages soil-living creatures like worms to do the spadework for you.
2 It reduces the loss of moisture.
3 It protects the soil structure.
4 It prevents weed seeds being brought to the surface.
5 It's easier on the back.

Five reasons to dig your plot

1 Digging breaks up heavily compacted soil, allowing it to breathe.
2 It kills surface weeds.
3 It exposes pests to predators and the cold.
4 If you don't you'll need a lot more mulch to grow potatoes as you will not be 'earthing up'.
5 It's good exercise for those who want it.

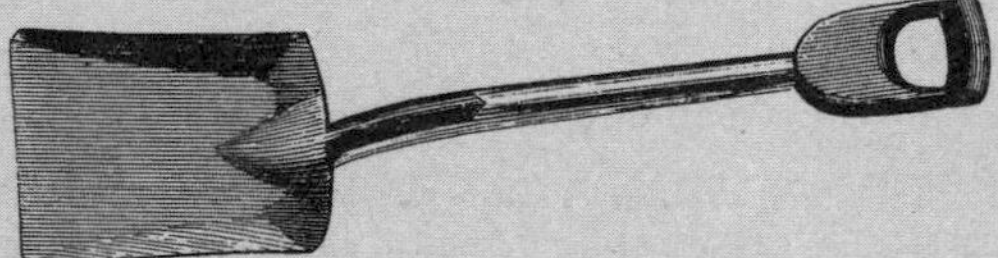

The **wild parsnip** can be found growing throughout Britain in overgrown areas such as roadsides and near railway tracks, but it is no good for eating. The juice from its stems and leaves can also cause people with sensitive skins to develop a burning rash. The yellow umbellifer flowers are similar to those of the fennel.

Chilli plants will often produce far too many peppers for immediate use, but you can preserve them either by making them into a paste (chop them and then fry lightly in some olive oil) or by freezing them. For the latter, cut them open and remove all the pith and seeds, blanch for a minute in boiling water and then bag them up. If you are lucky enough to have a surfeit of larger peppers (the capsicum varieties), you can freeze them in the same way after blanching them for a minute or two longer.

Traditional country wisdom dictates that the **best time for**

planting is in the late afternoon or **early evening**, when the sun has cooled.

Mushrooms need high-quality soil to thrive. In the past, growers used to use the soil from molehills because it is rich in nutrients, having been delved from deep under the ground.

What a man needs in gardening is a cast iron back with a hinge on it.

CHARLES DUDLEY WARNER (1829–1901)

A VAST GREENHOUSE COMPLEX, the size of **20 football pitches**, has recently been created in Billingham, near Middlesbrough, to grow over 200,000 tomato plants all year round. It is unique for Britain in that it **runs on waste** from a nearby factory. Steam and carbon dioxide, which would otherwise be released into the atmosphere, are pumped in to provide heat and boost growth. More than 5,000 bees have been introduced to pollinate the plants and it is hoped that the complex will produce over 7,000 tons of tomatoes every year to satisfy the increasing demand for British vegetables grown in the winter months.

The cucumber is a native of India.

THE **TOMATO** WAS ONCE KNOWN as 'wolf peach' on account of its allegedly dangerous properties. (Its resemblance to deadly nightshade deterred people from eating it.) US president Thomas Jefferson did much to help the gradual acceptance of the tomato as a nutritious and tasty vegetable, but it was another American, Robert Gibbon Johnson, who did most to convince people of its harmlessness. In 1820 he ate an entire basket of tomatoes at a public gathering in Salem, New Jersey, to the horror of his audience, but when he survived, the popularity of the tomato quickly began to spread. Today, it is one of the world's best-loved vegetables, grown from the tropics to the sub-polar region.

BROCCOLI WAS INTRODUCED TO the United States in the 1920s by Italian immigrants arriving in New York.

THERE IS A LITTLE BIT OF TRUTH in the old belief that **carrots** enable us to see better in the dark. The orange beta-carotene is a significant source of vitamin A, which is needed for the optic nerves, but in truth vitamin A is also important for the healthy functioning of many other parts of our bodies. This semi-myth, however, did play a role during the Second World War in helping the Royal Air Force to disguise from the Germans its use of the newly introduced radar system. Stories were fed to the British public that the superior skill of our **Spitfire and Hurricane pilots**, particularly in night-time sorties, was largely down to the vast amount of carrots that they were eating. The disinformation was extensively reported throughout the British press and it is probably responsible for the widespread belief held ever since that carrots are good for the eyesight.

FOR MANY CENTURIES IN Western Europe, the **aubergine** was thought to be deadly to Christians.

THE LATIN WORD FOR THE **chickpea** is *cicer*. The orator and statesman Cicero took his name from it because one of his ancestors was thought to have had a facial wart that bore a close resemblance to a chickpea. Try to forget that fact the next time you tuck into a chickpea salad.

IF YOU ARE ALWAYS SHORT OF kindling for your fire, keep the stalks of **sweetcorn** plants and dry them out. They make excellent firelighters.

On spinach:
Everyone knows the quality of this excellent plant. Pigs, who are excellent judges of the relative qualities of vegetables, will leave cabbages for lettuces, and lettuces for spinage.

WILLIAM COBBETT
(1763–1835)
The English Gardener

Eat leeks in March
and garlic in May
And all the year
the doctors play.

OLD SAYING

'CLAMPING' IS A USEFUL WAY TO store root vegetables such as potatoes, turnips, swedes and beetroot because most of us do not have the space indoors to keep large piles or boxes sitting around. The process was highly recommended by the Ministry of Agriculture during the Second World War, when the government was very keen that we grew and preserved as much of our own food as possible. All you need is some straw, a bit of earth and a dry spot in the garden. Once the freshly dug vegetables have been allowed to dry out for an hour or two, place them pyramid-style in a straw 'sandwich' with some underneath and some on top, and add a few spades' worth of dry soil to stop the straw from blowing away. Let them perspire for a couple of days and then completely cover the mound with soil. A well-built mound will keep your vegetables in good condition for six or seven months (i.e. deep into winter), but it is worth checking every now and then to remove any rotten ones.

Carrots and beetroot can be stored in boxes in layers of sand, sawdust, dry soil or even fire ashes.

TO PICKLE BEETROOT, BOIL for about an hour and a half (longer if they are large) and rub off the skins, which should surrender quite easily. Chop or slice them before pickling them in jars (see p. 49). You can start eating them after a week or two but one of the world's leading experts (me) suggests you resist them for about 10 weeks. New Zealanders like eating them in their hamburgers or 'Kiwiburgers'.

The greatest delight which the fields and woods minister is the suggestion of an occult relation between man and the vegetable.

RALPH WALDO EMERSON
(1803–82)

The cucumber was cherished by the ancient Egyptians, who made a drink from it by cutting a hole in one end and stirring up the flesh inside with a stick.

ALLOTMENTS BOOMED IN BOTH world wars, following calls from the government for families to grow as much of their own food as possible. Britons were also encouraged to rip out their flowers and shrubs and convert their gardens into vegetable and fruit plots, while many public parks were also turned into market gardens. In the First World War, George V led by example by replacing his flower beds at the front of Buckingham Palace with potatoes. Over 1.5 million allotments were in use during the Second World War. Today there are roughly 300,000, but interest in them is growing again in the wake of food scares and greater demand for organic produce. They can be rented for as little as £5 a year. Ask your local council or visit the website for the National Society of Allotment and Leisure Gardeners, www.nsalg.demon.co.uk.

THE **AUBERGINE** WAS PROBABLY first consumed in China around AD 500. Two-thirds of the world's aubergines are grown in the American state of New Jersey. Americans and Australians call the aubergine the eggplant because many of the earliest varieties were white and resembled hen's eggs.

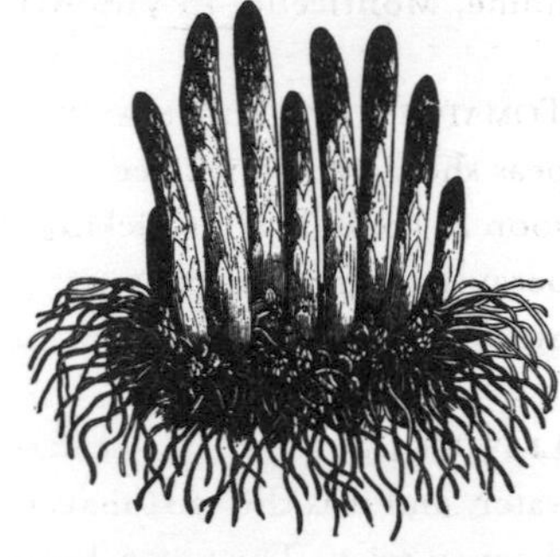

BATTERSEA PARK WAS ONCE THE site of a giant **asparagus** field, with over 260 acres set aside for its cultivation.

Of all the intolerable bores who visit us is the man who brings his own place with him and who, whatever may be shown him, at once institutes a comparison with his own, and begins to tell that 'mine are much better than that – I can beat you on so and so,' and ignoring the thing before him tells us 'Ah you should see my strawberries', 'my roses', 'my tomatoes', and so all through – in short the man who does not 'shut his own gate behind him' . . . From the chronic boaster of his own achievements we hope to be delivered.

JANE LOUDON
The Amateur Gardener, 1847

Thomas Jefferson, US president between 1801 and 1809 and author of the Declaration of Independence, was an avid gardener and grew 12 varieties of lettuce at his home, Monticello, in Virginia.

Tomatoes, sweetcorn and peas should be consumed as soon as possible after picking because their **sugar content** quickly decreases.

Lettuce leaves are 95 per cent water, and it is this that makes them so crisp. The water-heavy cells are tightly packed against each other, producing the crunchy texture. For scientific reasons too complicated and dull to go into (it's to do with osmosis), you can restore some of the crispness in your limp lettuce leaves by submerging them in water.

Sex is good but not as good as fresh, sweet corn.

Garrison Keillor (b. 1942)

If your potatoes happen to have caught the sunlight and **turned green**, you can still derive benefit from them by boiling them up and using the water, when cooled, as an effective insect repellent for spraying on plants.

Hops are vigorous, fast-growing plants that are attractive and useful if grown over a structure in your garden such as a shed or a gazebo. (Bear in mind that hops are not mad about wind, so avoid an exposed site.) You can dry the light green flowers and use them to make a hop pillow (see p. 103), but if you fancy a good, honest, **no-nonsense old-fashioned beer** that will be ready to quaff in roughly a fortnight, read on. Dry out the hops by hanging them in the house (they look quite nice indoors) or in your shed. Put 1.5–2oz dried hops into a pan with 8 or 9 pints of water and boil for half an hour. Let the water cool until it's warm to the touch, and at the same time place a jar with 2lb malt extract in a pan of hot water. Once it has loosened up, pour the malt, together with 1½lb of sugar, into a container and then add the hop water which you have just strained. Finally, mix up three teaspoons of dried yeast with a teaspoon of sugar in some warm water and add it to the container. Give it a stir before sealing the container. After ten days pour the liquid into bottles to within a couple of inches of the top, add a teaspoon of sugar to each bottle, seal and leave for four or five days.

FREEZE **SPINACH** BY BLANCHING it for a minute or two and then bagging up after pressing out most of the water. It's best to freeze spinach in the size portions that you are accustomed to using because otherwise it will stick together in a great green block and be difficult to separate without **hacking it to pieces**.

THE **ASPARAGUS SEASON** IN England lasts for only six weeks, starting in early May. Traditional wisdom has it that we should stop picking the spears on the longest day, 21 June, the official start of summer, so that the plants can begin to garner their energy for next year's crop.

A FEW WEEKS BEFORE YOU plan to sow your **courgettes**, put the seeds in your pocket and allow them to get a little roughed up. This will help germination. If you forget, or you find it annoying to have seeds mingling with your coins, then you can scrape them with a rough stone.

SPICY COURGETTE FRITTERS

Slice up some courgettes into discs and dip them in plain flour flavoured with cumin powder and paprika ✿ Heat some olive oil in a frying pan, put in the floured courgettes and cook until brown and crispy on both sides.

MANY SEASONED GARDENERS suggest toughening up your **young seedlings** in the greenhouse by running your hands over them as often as a dozen times a day. Seedlings reared in the pampered, cosy world of the glasshouse are never as robust as their outdoor, streetwise counterparts, but this old trick is a good way of preparing them for life in the outside world.

SLUGS AND SNAILS CANNOT abide sharp, uneven surfaces, so it's a good idea to scatter the area you want to protect with grit, sharp sand, thorns or crumbled eggshells. The slimy pests are not fond of ash or soot either. Some gardeners prefer **diversionary tactics**, tempting the slugs away from plants by creating a pile of alternative goodies for them, such as old lettuce leaves and other vegetable matter, left close to the plot. Before retiring for the night, gather up the pests and, if you're humane, relocate them – perhaps in the garden of a neighbour with whom you have recently fallen out. Alternatively, you could just kill them in any way you see fit (a bucket of salt water does the trick nicely). Another humane way of trapping them is to lay some large leaves (rhubarb, pumpkin, courgette, etc.) near your plants. In the morning you will probably find a number of them have opted to rest up under the leaves after a hard night's chomping in this conveniently situated, dark, damp accommodation. Again, deal with the critters as you deem appropriate.

I incline to the opinion that we should try seeds as our ancestors tried witches, not by fire but by water and that . . . we should reprobate and destroy all that do not readily sink . . .

WILLIAM COBBETT (1763–1835)

Shallots should be planted on the shortest day of the year and harvested on the longest.

OLD SAYING

IF YOU PREFER YOUR **CUCUMBERS straight**, there is an easy way to achieve this. Buy some clear plastic tubing the width of a good-size cucumber, cut it to cucumber length and hang it beneath the fruit by lacing a thread through two holes at the top of the tube. The cucumber will then grow into the straight tube.

Never sow vegetable seeds too deep because it forces the young plants to struggle their way to the surface, depleting them of the energy they will need to grow and thrive above the surface.

CATERPILLARS CAN DEVASTATE A crop of **cabbages** like a biblical plague of locusts, but you can help protect younger plants by adding about half a tablespoon of salt to your watering can. The plants absorb the salt, making them unpleasant for caterpillars who bloat and die if they consume it, but the cabbages will have expunged the salt by the time they reach maturity.

The heads of calabrese, to give the broccoli we eat its real name, are made up of thousands of tiny unopened flowers. Calabrese means 'from Calabria', a region of south-west Italy. Broccoli means 'little shoots'.

IN THE LATE 1700S, WHEN Frenchmen went to war with virtually anyone in Europe who so much as looked at their girlfriends or spilled their lagers, Napoleon offered a prize of 12,000 francs to anyone who could come up with a method for **preserving food**. Healthy daily rations were essential to feed a huge army on the march. It was over 10 years before Nicolas Appert, a confectioner now known as the 'father of canning', scooped the award for his method of preserving food in airtight glass bottles.

BY 1818, OVER **FIVE MILLION acres** of British land, once accessible to working people for the growing of their own produce, had been 'enclosed' by powerful landowners in the wake of parliamentary legislation. Historians claim that the enclosures were one of the principal causes of the poverty and disease that blighted the country by the middle of the century. By mid-century, new legislation had been passed by MPs to address the problem, allowing for the development of '**field gardens**' (allotments to you and me) for the poor, after it was recognized that fresh, homegrown produce and exercise in the open air brought health benefits.

IN MAY 1917, SHORTLY AFTER the United States entered the First World War, President **Woodrow Wilson** wrote an open letter to the American people, published in *The Garden*

Magazine, calling on them to help the war effort by growing their own: *Every one who creates or cultivates a garden helps, and helps greatly, to solve the problem of the feeding of the nations . . . every housewife who practices strict economy puts herself in the ranks of those who serve the nation. This is the time for America to correct her unpardonable fault of wastefulness and extravagance ... Without abundant food, alike for the armies and peoples now at war, the whole great enterprise upon which we have embarked will break down and fail.*

Horseradish will spread through your garden like bushfire if you give it half a chance, so it is best to grow it in containers. Making fresh horseradish sauce is almost laughably simple, unless you happen to get any of the vapours in your eyes. (Horseradish in the eye makes onions feel like a cooling eyebath. You will run screaming from the kitchen as if your hair is on fire if this happens, so wear glasses, stand as far back as possible and wash your hands immediately. To be on the safe side, make your meddling children retreat 20 paces.) Grate the horseradish and add to a bowl with natural yoghurt, cream or even mayonnaise. That's it. Honestly. You can also create the same creamy consistency by mixing the grated horseradish with mustard powder, sugar and wine vinegar, but it's best to experiment according to your taste. Some like it hot, others mild.

If the insects in your garden are not going about their business as usual, courgettes may need help in pollinating. Take a small, clean artist's brush and rub it over first the male flower and then the female, which is distinguishable by the embryonic fruit at the base of the bloom.

Our vegetable garden is coming along well, with radishes and beans up and we are less worried about revolution than we used to be.

E. B. White (1899–1985)

Soil needs to breathe. To avoid compacting the earth in a vegetable plot without stepping stones or paths, walk along an **old plank** laid down between rows so that your weight is spread across a greater area, causing minimum damage.

More than 1,000 varieties of **tomatoes** are currently being grown in the US.

In 1893 a decision in the United States **Supreme Court** declared that the tomato was a vegetable and not the fruit which, in strict botanical terms, it is. The ruling had nothing to do with rigorous scientific enquiry and everything to do with money. The original case was brought against a **tomato** importer called John Nix, who had been refusing to pay the 10 per cent tariff levied on all imported vegetables on the entirely reasonable grounds that the tomato was not a vegetable.

You can freeze French beans just as you would any other green vegetable by blanching them and bagging them, but if it's **haricot beans** you're after, wait until the pods have yellowed and then pull up the entire plant and hang it up somewhere dry like the shed or the garage. (They don't look great in the sitting room.) Once the pods have stiffened, shell the beans and spread them out on a tray for a couple of days before transferring them to jars or other airtight containers.

Squidgy, **overripe tomatoes** can be rescued by placing them in a bowl of salted water for about 20 minutes.

10 vegetables suitable for container growing

Aubergine	Horseradish	Radish
Carrots	Lettuce	Tomatoes
Courgettes	Parsnips	
Cucumber	Potatoes	

To prevent water running off the compacted, heavily rooted soil of your grow-bags, when you first put in your tomatoes or other plants, take **an old plastic water bottle**, cut off the bottom part of it and bury the narrow end in the soil. Water the plants through the bottle. Plastic water bottles, cut off at the bottom, also make first-rate propagators for seeds and young plants in small pots.

Traditional country wisdom recommends that seeds be left

to soak overnight in a '**manure tea**' to encourage rapid germination. Fill a bucket with equal amounts of water and manure and leave it for a day before straining and diluting it until it is the colour of tea. Then pour the 'tea' into smaller cups and leave the seeds to soak for a day or night.

TERRACOTTA POTS ABSORB WATER readily, drying out the earth inside more quickly than pots made from other materials. To counter this problem, line the inside of the pots with **shopping bags**, which have been holed for drainage at the bottom. Some gardeners recommend soaking the pots before planting so that they don't take away the moisture vital to young plants.

THERE IS SOUND THINKING behind the old saying that garlic should be planted on the shortest day and harvested on the longest. **Garlic** needs a cold period, preferably a sharp frost or two, for it to grow vigorously. In fact, you don't have to wait until 21 December to plant your cloves; and any time from mid-autumn onwards is fine.

A FAKE SNAKE MAKES AN excellent bird deterrent on a vegetable plot. Use one of your children's old rubber ones, or, just as good, cut up some old hosepipe and bend it once or twice into a snake shape. Make sure you place it somewhere visible (i.e not under foliage), and move it around occasionally. Birds may have brains the size of marbles, but they soon cotton on.

A SIMPLE WAY OF STOPPING your **lettuces** from bolting in the heat is to dig them up, put them in a cool place for an hour or so and then replant them. The ordeal of eradication is so upsetting for the plants that they shut down and stop growing.

TOMATOES WERE FIRST BROUGHT to England in the mid-sixteenth century from Central America via Italy (where the plant is still known as *pomodoro* or 'golden apple'), but it was not for another three and a half centuries that British people began to eat them. The leaves were similar to those of deadly nightshade and cattle refused to eat the bushes, and so they were ignored by British cooks until the twentieth century.

Garden Wildlife

Grass is hard and lumpy and damp, and full of dreadful black insects

ONCE **SWALLOWS** HAVE LEFT Britain for Africa at the onset of our winter, most choose to cross the Mediterranean at its narrowest point, near Gibraltar.

When swallows fly low, rain is on the way.

This old saying stands up to scientific inspection: when the air pressure falls and the air is moist, airborne insects descend towards the ground, hotly pursued by the swallows and swifts eager to eat them.

THE **ROBIN**, BRITAIN'S NATIONAL bird, is savagely territorial. The males will fight each other, sometimes to the death. The only garden bird to sing throughout the winter, it can be 'tamed' to some extent, but bird-lovers advise against this as it makes robins more vulnerable to cats and other animals.

HOUSE SPARROWS ARE A COMMON sight in the gardens of the **Falklands**. The birds colonized the islands after travelling to the South Atlantic aboard a fleet of whalers. A colony of sparrows once set up home in a Yorkshire colliery, 700 feet below ground, and lived on food brought to them by the miners.

BRITAIN HAS AN AVERAGE OF one magpie and one dunnock per garden. Both birds are sedentary and **rarely move** more than a kilometre from where they were hatched.

What would become of the garden if the gardener treated all the weeds and slugs and trespassers as he would like to be treated?

T. H. HUXLEY (1825–95)
British biologist, known as 'Darwin's Bulldog' for his ferocious defence of Darwin's Theory of Evolution

WHEN A **HONEYBEE** DISCOVERS a new source of nectar, she returns to the hive to announce the good news by carrying out a kind of **mid-air dance** through which she is able to communicate the location of her find. A worker bee will gather no more than one-tenth of a teaspoon of honey in her entire life.

A swarm of bees in May is worth a load of hay
A swarm of bees in June is worth a silver spoon
A swarm of bees in July is not worth a fly.

OLD SAYING

PRISTINE WEED-FREE LAWNS may look great but they are of little interest to most wildlife. So the next time your grumpy old uncle or a **snotty neighbour** pulls rank on the state of your lawn, remind them that your grass is home to a far better class of wildlife than theirs. Blotchier, more dishevelled lawns support many more species of beneficial insects than immaculately lined, shaved and trimmed ones of uniform colour. Even the more unwelcome insects serve a purpose by attracting creatures from higher up the food chain, such as birds and hedgehogs. Some people get obsessive about clover, but it's actually doing the garden a power of good by feeding it with nitrogen.

IF THERE WERE **NO BEES**, 100,000 plant species would cease to exist.

THE MAJORITY OF BLACKBIRDS (*Turdus merula*) live permanently in the United Kingdom, but about 25 per cent prefer to winter in France or Ireland.

THE **HONEYBEE KILLS** MORE people around the world each year than all the poisonous snakes combined, but the creature responsible for the most human deaths worldwide is the mosquito, by a considerable margin.

FROGS AND TOADS ARE A relatively rare sight in most of our gardens these days, but curiously it is in the countryside that their numbers have fallen most heavily, owing to, among other factors, the dredging of ponds for farmland, the loss of hedgerows and the use of pesticides. Frogs breathe through their skin and are highly **sensitive to pollution** in the water. Thus environmentalists tell us that the recent steep decline of the frog population indicates that we have very good reason to be worried about the world in which we now live. Like bees, ladybirds, hedgehogs, toads and most birds, frogs are a gardener's great friend because they eat so many pests, especially slugs and aphids. Creating a pond in your garden is the best way to attract frogs or toads. There is not the space here to explain exactly how to go about this but the Internet is full of good sites with straightforward instructions on how to make and maintain one. All I will say is that creating a pond is not particularly difficult and it will add a new, pleasant nook to your garden while attracting much beneficial wildlife. A pond is perhaps the most beneficial – to you and to wildlife – of all the habitats you can create in your garden.

IT IS ESTIMATED THAT **one-third of human food supplies** depends on pollination by insects, mostly bees. It is not known exactly how many bee species there are in the world but the number is thought to be around 30,000.

BEES ARE PARTICULARLY attracted to **blue, white, yellow and purple** colours for their

ultraviolet properties. It is no use planting red flowers in your garden to attract bees because bees cannot see red.

HERE'S SOME GOOD NEWS FOR the lazy gardener. If you have a **pile of wood** and/or branches sitting in a corner of your garden which you have been meaning to burn or clear for the past few months but haven't quite got round to it . . . then don't! Just leave it where it is because it will be teeming with wildlife, and most of the animals and insects that take refuge there will be doing a great deal of good work in the garden while you are sitting inside on your big bottom watching *Naked Dumper Truck Racing* from Pig's Town, Arkansas, live on Sky Sports 17.

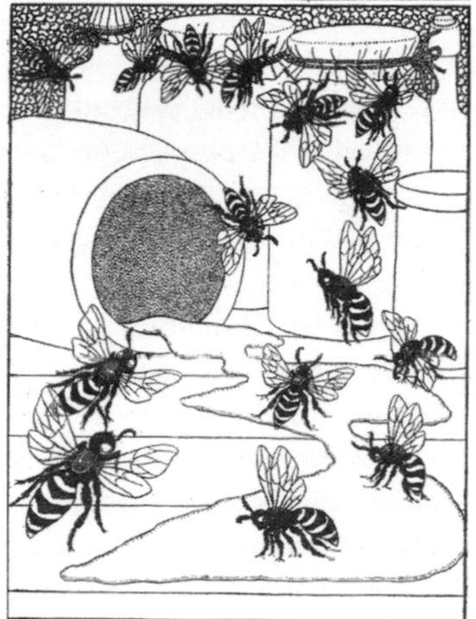

If the pile is in a relatively shady, damp, sheltered position then so much the better. Frogs, toads, hedgehogs, insect-eating beetles, bumblebees and centipedes will all gratefully make themselves at home in a wood pile.

10 plants honeybees love for their flowers

Dandelion
Fennel
Lavender
Nasturtium
Ox-eye daisy
Rosemary
Snapdragon
Sunflower
Teasel
Yarrow

10 plants bumblebees love for their flowers

Borage
Buddleia
Catnip
Comfrey
Foxglove
Geranium
Lupin
Sweet pea
Toadflax
White clover

On moles: *Put two or three heads of garlick, leeks or onions into their holes, and they will run out greatly terrified, so that they may then be easily caught by means of a dog.*

WILLIAM THOMPSON
The New Gardener's Calendar, 1779

BREAKING NEWS: **MOLES** ARE good for your garden! The soil they dig up and turn into hills may temporarily disfigure the quality of your lawn, but it is packed with nutrients, all perfectly crumbled, and can be scooped up and spread over your beds. If they are driving you mad, however, old-fashioned mothballs pushed into the holes are said to be a good deterrent. Moles also hate the smell of Jeyes fluid, which you can pour on to old cloths and stuff in. Others swear by planting empty bottles with the neck just above the ground – the noise of the wind in the bottle is almost deafening for the moles.

Nature is so uncomfortable. Grass is hard and lumpy and damp, and full of dreadful black insects. Why, even [William] *Morris's poorest workman could make you a more comfortable seat than the whole of Nature can.*

OSCAR WILDE (1854–1900)

A **MOLE** CAN SHIFT AROUND FIVE kilos of soil every 15 minutes, applying a force that is over 30 times its body weight. The mole has 44 teeth, which is more than any other mammal in Britain. Scientists know relatively little about moles, partly because they are elusive animals and spend their whole lives underground but also because they need so much food in captivity. They eat almost the equivalent of their body weight every day in worms, grubs and other insects. Moles are solitary animals who will fight each other to the death in a territorial dispute. When they are bracing themselves for a fight they let out a shriek that rises in a horrible crescendo.

A snail can sleep for three years. Ants don't sleep.

THERE ARE MORE INSECTS IN **one square mile** of rural land than there are human beings on the entire earth.

IN AN AVERAGE CUBIC METRE of soil you will find:
1,000,000,000 microbes
9,000,000 protozoa
5,000,000 nematodes
1,000,000 insects and arachnids
25,000 rotifers
25,000 worms
500 waterbears
100 slugs and snails
50 flatworms

Each year, insects eat one-third of the earth's food crops.

STARLINGS, AMONG OUR noisiest birds, make excellent mimics and often reproduce the sounds of other birds as well as frogs, animals and even machines! In addition to being a highly entertaining feature in your garden, starlings are very useful as they eat slugs, snails and leatherjackets. (Note to butterfly fans: starlings also love caterpillars.)

YOU CAN PROTECT YOUR POT plants from slug and **snail attack** by (a) smearing a band of Vaseline around the circumference of the pot or (b) wrapping around a length of double-sided sticky tape.

These are most anxious times on account of the slugs. Now every morning when I rise I go at once into the garden at four o'clock and make a business of slaughtering them till half past five, when I stop for breakfast.

CELIA THAXTER
An Island Garden, 1894

CATS KILL MOST WILDLIFE AT **dawn and dusk.**

Twenty plants for a meadow area in your garden

Ox-eye daisy, cat's ear, meadow buttercup, lady's smock, cowslip, St John's wort, bird's foot, yarrow, meadow sweet, lesser knapweed, wild carrot, poppy, meadow cranesbill, cornflower, red campion, field scabious, selfheal, bird's foot trefoil, musk mallow, toadflax

TO THE **SLOW WORM**, AS TO THE toad and the hedgehog, the slug is a tremendous delicacy, the fillet steak of the garden. These snake-like creatures are harmless to us and other mammals, but will gobble up a variety of insects. (In spite of their name, they are actually fairly fast over the ground, especially if a cat or a passing bird of prey makes a move for them.) They like humid habitats, living under rocks and log piles, and on a hot day will often be found sunning themselves on the compost heap.

IF YOU HAPPEN TO HAVE A **scruffy, overgrown corner** to your garden, perhaps full of nettles and brambles, which you have been meaning to do away with, you may be pleased to know that it's a very good idea to leave it exactly as it is. This unpromising area will be providing a home or a source of food to a wealth of butterflies and insects (which in turn will attract even more welcome visitors to feed on them) as well as birds and animals like hedgehogs. But if you continue to find the daily sight of this dishevelled area a terrible affront to your aesthetic sensibilities, still think twice before getting rid of it because there are alternatives. You can create a natural screen with some hedging (which will attract even more wildlife) or sow some colourful meadow flowers around it to tart it up and draw the eye away from what lies behind.

A **FREQUENTLY CLIPPED HEDGE** brings many benefits to the wildlife in your garden. The berry-producing plants provide food while the debris of fallen leaves and branches offers the perfect habitat for slug-hungry hedgehogs. A dense hedge will also act as a natural 'corridor' for wildlife to move along freely. Avoid clipping hedges in the nesting season of spring and early summer.

HEDGEHOGS ARE GOOD runners, climbers and swimmers but they often drown if they fall into a pool of water because they exhaust themselves trying to get out.

WORM CASTINGS, aka **worm shit**, contain high levels of potassium, nitrogen and phosphorus – the very elements that farmers spread all over their land and gardeners buy in manufactured forms from garden centres.

IT MAY SEEM STRANGE BUT THE **more birds** you have in your garden, the safer it is for them because there are that many more beady eyes looking out for the approach of cats.

I would not enter on my list of friends the man who needlessly sets foot upon a worm.

WILLIAM COWPER (1666–1709)

10 ways to prevent or treat aphid damage

1 Encourage predators such as ladybirds, spiders and hoverflies.
2 Go easy with the nitrogen fertilizer as the softer growth it encourages attracts aphids.
3 Inspect vulnerable plants regularly and break off affected shoots and leaves.
4 Plant pot marigolds close to the plants you want to protect.
5 Spray affected plants with derris or very diluted washing up-liquid.
6 Broad beans, a popular host plant for blackfly aphids, are less susceptible to attack if sown in autumn.
7 Hang up strips of fat in winter to attract blue tits (which eat aphid eggs).
8 Use resistant plant varieties where possible.
9 Snap off top shoots of broad beans as soon as they reach maximum height.
10 Burn or bury heavily damaged plants.

EARTHWORMS ARE USED BY research scientists looking to improve human medical conditions because their bodies have many similarities with our own: nervous system, blood vessels, haemoglobin, kidney-like organs producing urine . . . But don't get too worried about the weird relations you never knew you had because worms also have five hearts and both male and female reproductive organs, they breathe through their skins and when they want to eat they stick their throats out of their mouths to grab their food. It's going to be a while before they start moving into houses and driving cars.

EACH GARDEN contains an average of 100 species of **spiders** and each house contains about 10 species.

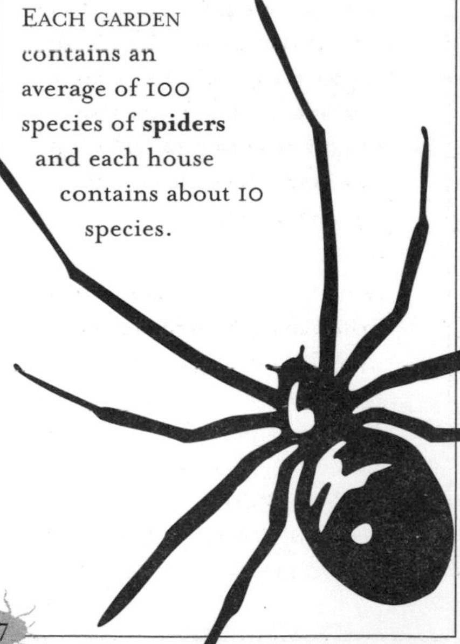

IN A PAPER WRITTEN LATE IN his life, **Charles Darwin** claimed that almost all the world's soil had at some point passed through the gut of an earthworm. Subsequent scientific studies have shown that his calculation was a little far-fetched, but he was thinking along the right lines.

A FEW plants need **moths** to pollinate them and they give off a powerful scent at dusk in order to attract these mainly nocturnal creatures. By day these plants barely smell at all because they have no good reason to draw attention to themselves. Common garden plants that attract moths include red campion, red valerian, sedum, hebe, nicotiana and the sallow tree.

HONEYBEE SOCIETY IS HIGHLY sophisticated and well organized. In a colony of bees (containing anything from 20,000 to 80,000 insects) a two-day-old larva is selected by the workers (female) to be reared as the **queen**. The drones (male) have one task on earth: to mate with the queen, leaving their genital equipment inside, and then drop dead. (It's better than working for a living, I guess.) The rest of the drones are expelled in the autumn because they are of no further use to the colony. The queen emerges from her cell after nearly two weeks to mate in flight with the drones. She starts to lay eggs about 10 days after mating and can produce up to 2,000 a day. She lays her own weight in eggs every couple of hours and is surrounded by workers, who **attend to her every wish**, giving her food and disposing of her waste. They also lick her body for the pheromones that are needed for the well-being of the colony. Of 100 bees, one is the queen, about 15 are drones and the rest are female workers. A queen bee can live for up to three years, but workers die within four to six weeks. The entrance to the hive is constantly manned by guard bees, the **bouncers of the insect world**, who can detect an intruder by their different smell. The guards organize an all-out attack on the interloper, the dead body of which is then removed from the hive by the 'undertaker' bees, whose primary task is to get rid of the dozens, even hundreds, of their fellow worker bees who pass away each day.

TO ANIMALS, **THE SMELL OF elder leaves** is so awful that even rabbits won't eat them. In days gone by gardeners used to place a few branches of elder in their gooseberry bushes to stop magpie moths stripping off all the leaves.

Aphids are among the most damaging pests in the garden. There are over 500 species of them in England. They feed by sucking the sap out of plants and it is by them that many plant viruses are spread.

MICE LOVE EATING THE FRESHLY sown seeds of peas and beans. There are a number of ways to avoid this happening. If you don't have a greenhouse, sow them as normal and then cover the rows with fine chicken mesh and peg it into the earth. Alternatively, if you have plenty of prickly twigs and branches available, scatter them fairly densely over the row. Another simple, ingenious method for those with a greenhouse is to start the seeds in a length of plastic guttering cut to size. When the plants are about two inches high, transplant them outside by sliding the row of peas and the soil straight into a prepared trench.

Bullfinches and tits do considerable harm to currants and gooseberry trees by destroying the buds; this harm is best combated by twisting black cotton in and out among the branches, rather than by killing these songsters and damaging the trees with shot.

GEOFFREY HENSLOW
The Gardener's Calendar
1925

Only female mosquitoes bite and drink your blood because they need the protein from it in order to produce eggs. Male mosquitoes, meanwhile, have no interest in you whatsoever, preferring to feed on nectar and water. Mosquito 'repellent' is a bit of a misnomer because all the spray does is jam the mosquito's sensors so that it is unaware that you're there.

WITH FARMING **practices** being what they are today, gardeners, should they wish to, can play a significant role in preserving our natural heritage and boosting the numbers of many animals, insects and birds. There is a vested interest in doing so too, because encouraging wildlife is good for the garden. The population of British bumblebees and honeybees, for instance, has dropped by roughly 50 per cent in the last 50 years, mainly because of modern agricultural practices, the use of pesticides and loss of habitat. Many farmers, however, have come to recognize the importance of bumblebees as pollinators and have started to cultivate them for that purpose. Although bumblebees can sting, they are one of the more docile and harmless creatures in your garden. They often live in holes in the ground vacated by rodents, but compared to those of honeybees their colonies are small, with no more than about 50 living together at a time. You can encourage bumblebees to live in your garden in a number of ways. You can buy or make **bumblebee boxes**, which are the size of shoeboxes and have two rooms: one for the queen and one for all her minions. They are also happy to live in a pile of logs or rocks. If you decide to put a box in your garden, fill it with hay or straw and put it in a sheltered, south-facing site out of direct sunlight, close to pollen sources such as a flower bed or hedge.

Some creatures in your garden are both enemy and ally. The **blackbird**, for instance, will help rid your garden of soil grubs and caterpillars, but it will also happily devour your strawberries and other soft fruits. The best predators of slugs are frogs, toads and hedgehogs, but the first two really need a small pond to establish themselves. Hedgehogs are voracious eaters of many undesirable insects, and will often travel over a mile to forage for their food.

Ninety-five per cent of Britain's **wild flower meadows** have disappeared in the last 50 years, with drastic consequences for much of our wildlife. Some environmentalists, far more noble and determined than the rest of us, seek to help the environment by abseiling down Big Ben or North Sea oilrigs, unfurling giant banners reading '*Save the Gay Hedgehog! Now! . . . Behead All Farmers! Now!*' but you can do your bit without troubling the constabulary or risking your neck. Simply plant a small meadow area in your garden. It may not be that dramatic and it won't get you on the *10 O'Clock News*, but dozens of insects, birds and animals will doff their caps in thanks and admiration each time you emerge into the garden thenceforth. A meadow will also look great and demand virtually no attention once it has established itself. Oddly perhaps, meadow flowers prefer poor soil, so if there's a rough area in your garden you will have a head start without lifting a finger. Even if you don't have a big garden in which to create a significant meadow area you can still create one or more small patches of meadowland to produce a similar effect. Bear in mind that most perennial meadow flowers planted by seed will take a year or two to establish themselves. Get rid of the grass and topsoil from the area, rake the soil to a fine tilth, sow the seed thinly and firm it down. In the first year cut the meadow several times to a height of about two inches, but never apply fertilizer.

It is something of an urban myth that a worm will be perfectly happy if you cut it in half. It may continue to wriggle for a while (so would you after you had been shot or stabbed), but it will die not long afterwards. Only if you nip off just a little of its tail end does it have the capacity to repair itself.

AS THEY RETURN TO THE NEST with food, ants lay down a **trail of pheromones** to attract and guide their comrades to the food source. The trail is continually freshened until the food runs out, when the pheromone evaporates quickly so that the ants are not confused when a new source of food is discovered.

WHEN **HUMMINGBIRDS** HOVER in front of flowers to gather nectar, they expend so much energy that humans would need to consume the equivalent of a 250lb steak to take on the requisite amount of calories to carry out a feat of similar effort.

THE **SKYLARK** POPULATION HAS been plummeting in recent years, but at the end of the nineteenth century the birds were so plentiful that every month thousands of them, mainly on the Sussex Downs and in East Anglia, were netted, killed and packed off to restaurants in Paris and London. Trading in skylarks has been banned in Britain, but the practice continues on the Continent.

GRASSES AND CEREALS ARE wind-pollinated plants so they don't have the colourful flowers or distinctive scents that other plants need in order to attract insects. Grasses and cereals produce an enormous amount of pollen, which explains why pollen allergy has come to be known as hay fever.

The honey-bee is not a native of our continent . . . The Indians concur with us in the tradition that it was brought from Europe; but when, and by whom, we know not. The bees have generally extended themselves into the country, a little in advance of the white settlers. The Indians, therefore, call them the white man's fly, and consider their approach as indicating the approach of the settlements of the whites.

THOMAS JEFFERSON
Notes on Virginia, 1782

VERY DRY OR FREEZING WEATHER can sometimes be good news for your vegetable plot and flower beds because earthworms have to dig deeper to find the high levels of moisture they need to survive. In doing so, they will be giving your soil a good airing as well as fertilizing it with their castings.

THE DEVIL'S COACH HORSE, a species of rove beetle found commonly in Britain, curls its tail upwards like a scorpion when threatened. It also boasts the ability to emit a **foul smell** from its anus and mouth at the same time. (A trick it almost certainly learned from an old university friend of mine.)

BREADCRUMBS ARE ACTUALLY OF little nutritional value to birds, although they are not harmful unless the bread contains a lot of salt.

NEVER PUT BIRD FOOD ON THE ground or on a low surface where it can be scavenged by foxes, rats and squirrels. If you want a **bird table**, make sure it is a high one and has been designed so that it is difficult for cats to pounce on birds. You can also attach defences, known as 'baffles', to the pole of the table to stop the squirrels climbing up. Don't put **bird feeders** on, or close to, low trees, bushes or hedges as cats will lick their paws at the sight of handy cover that will allow them to spring an ambush.

IF YOU HAVE deliberately set out to attract more birds to your garden, remember that they need water as much as they need food. Birds that eat mainly seeds, which are very dry, are in even greater need of water. Introducing a **birdbath** will help ensure that the birds stick to your garden rather than establish themselves in a place with a better water supply. Make sure the birdbath is positioned somewhere the birds will be able to see cats approaching.

Natural slug repellents

Gravel
Large bark chippings
Jars of salty water dug into the ground
Jars of diluted beer dug into the ground
Fresh seaweed (full of salt), but put it around the beds rather than on, as plants aren't fond of the stuff either
Toads and hedgehogs
Blackbirds, jays, thrushes, starlings and robins

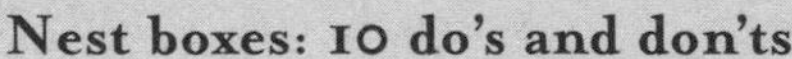

Nest boxes: 10 do's and don'ts

1 If you have introduced nest boxes to your garden, make sure that the lids are secure so that **squirrels and magpies** are unable to pillage the eggs.

2 Nest boxes should be placed somewhere out of the wind, rain and full sun (the **chicks will overheat** and die if the box is in the full glare of the midday sun).

3 Boxes should also be positioned where cats cannot get at them, but close to a tree or shrub with **small branches** so that the fledglings – but nothing larger – can perch on them.

4 If you make your own box, **don't add a perch** as larger birds may come and scare away the occupants.

5 If you want to make or buy a box for **blue tits or coal tits**, the hole should be roughly 2.5cm in diameter. For slightly larger birds like great tits and sparrows, it should be about 3.5cm.

6 Ensure your nest box is made from **untreated wood**.

7 Don't put any nesting material inside the box as birds prefer to do their own **interior designing**.

8 It is essential to **clean out** bird boxes with boiling water in the autumn, otherwise diseases and parasites will be passed on to new inhabitants.

9 The best time to put up a box is in the **early autumn**, which is the busiest season in the bird property market. Don't give up hope if your box remains uninhabited in the first year.

10 If you are planning to put out a number of boxes, don't place them too close together as birds like **space** they can call their own.

Not all slugs cause damage to your garden. The **great grey slug** is relatively harmless to your precious vegetables and flowers because it generally eats fungi and rotted vegetable matter and even its other, less welcome cousins. The European black slug has as many as 25,000 teeth. Although it will eat your prized plants, it serves a more welcome purpose by devouring dog and cat poo and turning it into fertilizer.

Although we do have a National Fancy Rat Society in Britain, someone is yet to found the Friends of the Wasp Society, largely because it is almost impossible to find anyone with a good word to say about these annoying parasites. **Wasps**, however, do a very modest amount of good in your garden by removing the caterpillars munching through your plants and feeding them to the larvae back in their nest – but this is not exactly a ringing endorsement because most people like to see butterflies in their garden, even if that means kissing goodbye to a few leaves of a cherished plant from time to time. Blackbirds and starlings love eating wasps, so stand in the middle of your garden and welcome these birds with open arms. They will also sing nice songs for you. If wasps are a real problem for you and your children, then the best way to deal with them is by employing some cunning diversionary tactics. Cut open some old pieces of fruit and place them as far away from the house as possible.

Most **insect bites and stings** contain high quantities of acid, so common scientific sense tells us that plants high in alkalis will work as an effective antidote. The best-known natural remedy is dock leaves but tomato and potato leaves and onion juice are just as good. Bicarbonate of soda mixed in water and applied to the affected area is another excellent antidote. **Wasp and hornet stings**, however, release alkalis into the skin and should be treated with something acidic like lemon juice or vinegar.

Wasps build their paper nests underground, and occasionally under the eaves of a building or even in an attic. The paper is produced when they chew wood into a pulp.

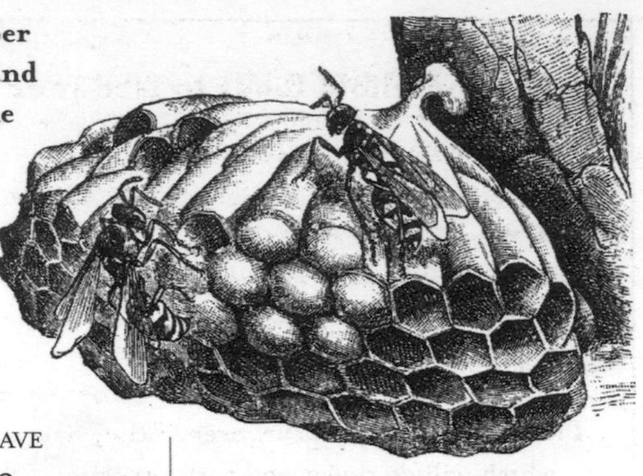

A HONEYBEE WOULD HAVE to fly around 100,000 kilometres and visit over a **million flowers** to find the nectar to make two kilos of honey. As it happens, they can only manage about 800 kilometres before they exhaust themselves and die.

AT THE END OF THE SUMMER you will often see small **clouds of flying ants**, particularly when a thunderstorm is brewing. These clouds are known as 'ant weddings' because this is when the male ants mate with the queen ants. Like honeybee drones, the males die after they have 'pleasured' the queen.

UNLESS I'M MISSING OUT ON something here, slug sex appears to be mightily different to human sex. The way slugs court is to circle each other for a while and produce a **great big puddle of slime**. Then, because they are hermaphrodite, they inject each other with sperm before slipping away to lay roughly three dozen eggs each.

WOODLICE ARE TERRESTRIAL crustaceans, distantly related to the lobster and the crab. Around 50 million years ago, for reasons best known to themselves, they emerged from the oceans for a life on land. Habit, however, dies hard in the woodlouse and they still prefer to set up home in a damp, dark environment because they need the moisture to breathe through their gills.

WRENS LIKE TO SLEEP together, especially in the colder months, and 46 of them were once found huddled up in a single nest box, according to the Royal Society for the Protection of Birds (RSPB).

In the last 50 years we have lost more than half the hedgerows in our countryside.

What household foods to feed your garden birds

Raisins, sultanas and currants
Fat, which is enjoyed by tits, thrushes and wrens
Cooked, unsalted rice
Grated mild cheddar cheese, a treat for robins, blackbirds, wrens and song thrushes
Dry porridge oats
Cold plain potatoes in all forms except chips
Pieces of apple and pear, even if they have started to go off, which will go down well with thrushes, tits and starlings
Pastry, cooked or uncooked
Peanuts sold for the express purpose of feeding to birds

What NOT to feed your garden birds

Cheap peanuts (they may contain a harmful toxin that can prove fatal to some birds)
Desiccated coconut (it can swell the stomach and kill some birds)
Spicy or salty food
Salted nuts, bacon, crisps and snacks
Margarine and vegetable oils
Food with mould on it
Chocolate

A ladybird can eat up to 150 aphids in a day.

THE **FOXGLOVE** flower has arranged itself to suit the demands of bees. The female flowers at the bottom of the stem contain the most nectar, which decreases in concentration the higher up the plant. By the time the bee has reached the male flowers at the top, the nectar has all but gone, thus persuading the bee to go to the bottom of another column of flowers. In doing so it very kindly brings the male pollen to the female flowers.

Birds operate at different levels in the garden. Wrens and blackbirds, for instance, enjoy vegetation on or close to the ground, finches prefer shrubs and hedges while the song thrush surveys all below from the treetops.

AT THE RISK OF ENCOURAGING you to turn your cat into a **one-animal orchestra** or a walking campanology society, wildlife experts recommend attaching more than one bell to its collar. Many cats have learned the art of moving so stealthily that their single bell won't even ring on approach.

IF YOU SEE **ANTS** CLIMBING UP A plant it is a fairly good indicator that aphids are at work because ants love the sugary substance that they secrete.

EVERY DAY earthworms **eat and excrete** their own weight in rotted vegetable matter and earth. They are no threat to plants because they only eat organic matter that has started to decay.

THE **DREADED WIREWORM**, which delights in destroying potatoes and other root crops, lives for three or four years underground, feasting on plant roots, before finally emerging as a click beetle. The insect takes its name from the noise it makes as it hurls itself into the air when it is attacked or touched.

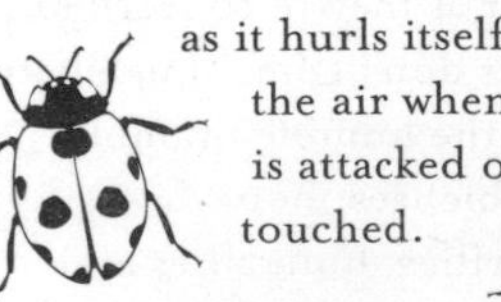

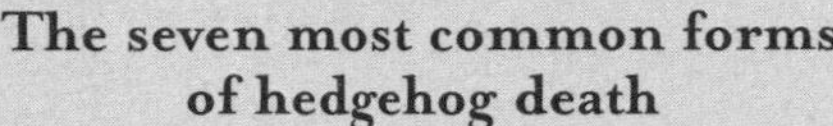

The seven most common forms of hedgehog death

1 Run over. (It's often the mothers hurrying home with food for their young.)

2 Poisoned by eating slugs that have consumed gardeners' slug pellets.

3 Trapped in a hole, ditch or pond.

4 By strimmers which kill them outright or cause injuries that kill them later.

5 Stabbed by a garden fork, often in a compost heap, where they go for warmth.

6 Burned at the stake. (Hedgehogs often live in unlit bonfires or woodpiles.)

7 Death by rubbish. Hedgehogs foraging in bins are vulnerable to sharp objects and sometimes get stuck inside tins and cartons.

BLACKBERRY BUSHES MAY HAVE had a sinister reputation in the Middle Ages (the devil was thought to urinate on them), but you try telling that to our insects, birds and mammals. As far as they're concerned, plants don't come much better than the humble bramble. Bumblebees, honeybees, hoverflies, butterflies and moths flock to its flowers for pollen and nectar, while blackbirds, thrushes, robins and starlings are just some of the many birds that enjoy the fruit. Small animals – even foxes – are also partial to a blackberry. Many birds and small mammals like to nest in its tangled bushes, where they are safe from larger predators.

SUMMER

In June, as many as a dozen species may burst their buds on a single day. No man can heed all of these anniversaries; no man can ignore all of them.

Aldo Leopold (n.d.)

People don't notice whether it's winter or summer when they're happy.

Anton Chekhov (1860–1904)

Heat, ma'am! It was so dreadful here that I found there was nothing left for it but to take off my flesh and sit in my bones.

Sydney Smith (1771–1845)
writer, humorist, clergyman

Summer makes a silence after spring.

Vita Sackville-West (1892–1962)

The trees that have it in their pent-up buds
To darken nature and be summer woods

Robert Frost
'Spring Pools', 1928

A life without love is like a year without summer.

Swedish proverb

Summer Birthdays

18 July 1720: Gilbert White, Selborne, Hampshire
3 August 1801: Joseph Paxton, Woburn, Bedfordshire
4 August 1608: John Tradescant the younger, Meopham, Kent

To see the Summer Sky
Is Poetry, though never in a Book
it lie
True Poems flee.

EMILY DICKINSON (1830–86)

In the depth of winter, I finally learned that within me there lay an invincible summer.

ALBERT CAMUS (1913–60)

Summer vegetables

Artichokes, aubergines, beetroot (early), broad beans, broccoli, lettuce, carrots (baby), courgettes, cucumber, French and runner beans, mangetout, peas, potatoes, Swiss chard

Summer afternoon, summer afternoon; to me those have always been the two most beautiful words in the English language.

HENRY JAMES (1843–1916)

Five increasingly rare summer visitors

Corncrake
Nightingale
Osprey
Spotted flycatcher
Turtle dove

What is one to say about June, the time of perfect young summer, the fulfilment of the promise of the earlier months, and with as yet no sign to remind one that its fresh young beauty will ever fade.

GERTRUDE JEKYLL (1843–1932)

Herbs

Herbs are the friend of the physician and the pride of cooks

In strict botanical terms a herb is a seed-bearing, non-woody plant that dies back to a rootstock in winter – a definition which rules out shrubs and plants like rosemary and thyme. For the purposes of this chapter, we'll use the old, wider meaning of any plant that is useful or **beneficial to man**.

For many centuries the vapours from mint infusions and other concoctions have been used for relieving congestion in the head and chest. Combined with elderflower and yarrow, mint is especially effective in relieving the **symptoms of colds**. Herbalists claim that rubbing in a few drops of mint oil or massaging the temples with some fresh leaves can ease tension headaches. Mint is also a good mental stimulant, similar to caffeine, but without making you feel jittery or racy.

Only **5 per cent** of all the world's plants have been tested for their potential medicinal properties, and yet 85 per cent of pharmaceutical drugs are linked to plants.

Good housewives in summer
will save their own seeds,
Against the next year,
as occasion needs.
One seed for another,
to make an exchange
With fellowlie neighbourhood
seemeth not strange.

Thomas Tusser
Five Hundred Points of Good Husbandry
1573

BASIL OIL IS delicious drizzled over sliced tomatoes, but it can also be used in cooking and as a general salad dressing. There are two simple ways of making it. First, whiz up the leaves with some olive oil in a food processor. Lightly heat the mixture in a pan for three or four minutes to release the flavour, and then let it cool before pouring into a jar. Alternatively, just put a large handful of the leaves in a jar of olive oil and allow it to rest for a month in a dark cupboard.

PINCH OUT THE TOPS OF YOUR basil plants to ensure future growth is outwards and not upwards. Don't let them flower either.

A certain gentleman of Siena being wonderfully taken and delighted with the Smell of Basil, was wont very frequently to take the powder of the dry herb, and snuff it up his Nose but in a short Time he turn'd mad and died; and his head being opened by surgeons, there was a Nest of Scorpions in his Brain.

JOSEPH PITTON DE TOURNEFORT (1656–1708)
The Compleat Herbal

There are no worthless herbs, only the lack of knowledge.

CHINESE SAYING

Six ways of using basil

- Add to scrambled eggs or omelettes
- Dry and make into powder
- Eat the leaves to aid digestion
- Rub into the skin as a mosquito repellent
- Rub into the temples to relieve tension headaches
- Sprinkle the leaves over a tomato salad

BASIL, A MEMBER of the mint family, is abundant throughout the tropical regions but can grow in cooler climates too, although in Britain it is best left in the greenhouse or indoors. Originally from India, basil (from the Greek word for king, *basileus*) worked its way through the Middle East and the Mediterranean before reaching Britain in the sixteenth century. In India basil has been used over the years to repel mosquitoes, which don't like the powerful smell of the estragol and eugenol it contains. Burn some basil on the barbecue to deter mosquitoes.

10 common ingredients in an Elizabethan salad

Bergamot
Chrysanthemum flowers
Dandelion
Fennel
Lovage
Mint
Nasturtium
Primrose
Purslane
Violet

What can be more pleasant to thee, than the enjoying of medicines for cure of thine infirmities, out of thy native soil, and country, thy field, thy orchard, thy garden?

NICHOLAS CULPEPER
The School of Physick, 1659

The pleasure and use of gardens were unknown to our great grandfathers: They were contented with pot herbs: and did mind chiefly their stables. But in the time of King Charles II gardening was much improved, and became common.

JOHN AUBREY (1625–97)

SEVERAL CENTURIES AFTER THE departure of the Romans from Britain, herb and vegetable gardens began to appear again, this time in the **monasteries** rather than the villas of the rich and powerful. Their purpose was twofold: to allow the monks to live self-sufficiently and to help them treat the sick and the infirm who came to them for help. By Elizabethan times all large houses had elaborate herb gardens, walled or hedged in from the weather. Most plants were grown for medicinal or culinary use but others, such as violets, roses and lavender, were sown for the making of cosmetics.

MINT AND APPLE JELLY

Take about a kilo of apples and cut them coarsely before putting them into a large pan with a little water and some lemon rind ❁ Simmer until they are soft and then bash them to a pulp while still cooking ❁ Put the mush into a jelly bag or a muslin cloth and suspend it over a bowl to drip overnight ❁ Don't squeeze the mixture or your jelly will be cloudy ❁ The next day for every litre of juice you pour back into the pan, add roughly 800g sugar and squeeze in the juice of a large lemon ❁ Boil it all up until it starts to set ❁ Depending on the pectin content of the apples, this should take between 15 and 30 minutes ❁ (You will not know if it is ready to set while it's still hot, so put a teaspoon of it on a saucer, and if a skin forms then it's ready to be lifted.) ❁ Add a cup of chopped mint and bring it back to the boil before leaving it to cool and transferring to jars that have been warmed so that they don't crack.

The Doctrine of Signatures, a medicinal theory popular in the early seventeenth century, held that certain plants could be used to cure human ailments according to their resemblance to parts of the human body. Thus the brain-like form of the walnut was prescribed for the treatment of headaches, the hairy roots of some plants were thought to help cure baldness and the yellow flowers of broom were believed to be good for jaundice.

Six ways of using rosemary

Add to apple jelly
Add to soups and stews
Dry and make into a powder
Make an infusion to clear a head cold
Put the leaves inside a chicken for roasting
Put a sprig in a glass of Pimm's

IF YOUR FEET ACHE AFTER A long day outdoors, try soaking them in a **mustard bath**. Make a tablespoon of mustard powder (by crushing the seeds) or, if you have some ready made, put it straight into a bowl of hot water and immerse your weary feet.

A decoction of the tops of [hops] *cleanses the blood, cures the venereal disease, and all kinds of scabs.*

NICHOLAS CULPEPER
The Complete Herbal, 1653

THERE ARE MANY SUPERSTITIONS surrounding **parsley**, that most British of herbs, probably as a result of its highly unpredictable germination. It is said to be very unlucky to transplant parsley, and there is an ancient saying that parsley goes to the devil and back seven times before germinating and that it will only thrive if planted by an honest man. As with many plants of the *Umbelliferae* family, parsley seed must be very fresh for sowing. Hot water poured on the soil half an hour beforehand will help its germination. Parsley is a biennial – prone to bolting in a hot summer in its first year – but it will become perennial if you stop it from flowering. 'Parsley bed' has long been used as a euphemism for female genitals and 'parsley' for pubic hair. Since the early 1700s parents have often told their children, inquisitive about how they made it into this world, that they emerged 'from the parsley bed'.

AN INGENIOUS AND HANDY WAY to preserve garden herbs such as basil, tarragon and chives is to freeze them, as soon after picking them as possible, in **ice**

Six ways of using parsley

Add to scrambled eggs
Chop into white sauce
Dry and add to homemade *bouquets garni*
Dry and make into a powder
Make parsley butter for basting chickens and serving with fish
Use in stuffing

cubes. When they have frozen solid, transfer the cubes into bags and put them back in the freezer. Larger-leafed herbs like mint and parsley can be frozen in bags as they are.

PESTO SAUCE

Take some fresh basil leaves, enough to fill a cup (you can also use rocket or flatleaf parsley if you have a profusion of either, or combine them with the basil), and put them in a blender with a few garlic cloves and a few tablespoons of olive oil ❁ Add a good handful of roasted pine nuts and another of grated Parmesan cheese with a bit of salt, and blend.

HERB GARDENS FOR **THE BLIND** have become increasingly popular in our towns and cities. Good strong-smelling plants include rosemary, lavender, tansy, lemon balm, bergamot and nicotiana. The domestic gardener may be tempted to site these plants where the fragrances can be most appreciated, i.e. in areas close to the house, such as by the front door, along paths and next to garden furniture.

Five common herb oils with antiseptic properties

Bergamot
Lavender
Lemon verbena
Rosemary
Thyme

LAD'S LOVE (common wormwood) is a bitter-tasting herb used in the making of the highly potent (and now illegal) drink absinthe. Van Gogh was said to have been drinking absinthe before deciding to cut off his ear. It is also known as *Artemisia absinthium*, which takes its name from the wife and sister (one and the same) of the Greek-Persian king Mausoleus.
For years it has been used as an insect repellent and it is particularly effective for seeing off moths. Some gardeners make a 'tea' from it and spray it over well-established (non-edible) plants to deter slugs and snails, but it's best not to use it on young seedlings because its toxicity will do more harm than good. To be on the safe side, you are probably better off drying some sprigs of lad's love and hanging them near the plants you want to protect.

NETTLES, COOKED AS GREENS, in soup or added to porridge, used to be a **common sight** on British tables. They have a very high mineral and vitamin content. **Nettle soup** is as delicious and nutritious as any green soup – or any soup for that matter. It leaves spinach soup for dead. The common nettle has fallen out of fashion in recent years, but it is long overdue a revival. This recipe is best **before the end of May** when the nettles start to get older and woodier. Using gardening gloves, collect roughly eight large handfuls of young nettle tops or a carrier bag of loosely packed leaves. Wash them. Melt a knob of butter and cook a large chopped onion with the nettles until they are soft. Add about 25g of flour and cook for a few more minutes, stirring all the time, before adding a litre of chicken or vegetable stock. Bring to the boil, simmer for five minutes and then sieve or give it a blast in a food processor. Reheat it before serving, season well and add a little cream or crème fraiche if you want, although it is perfectly delicious without.

The nettle is host to many of Britain's most **beautiful butterflies**, and environmentalists now encourage us to set aside a patch for them in our gardens. Modern farming practices, including the use of chemicals that drift into hedgerows and kill the plants there, have wiped out great swathes of nettles.

THE **ROMANS** BROUGHT THE seeds of one of our common nettles to help cure ailments resulting from the damper, cooler British climate. The legionaries were said to **whip themselves** with bunches of nettles to warm up their bodies and improve the circulation of the blood. Nettles used to be consumed in wine by more suspicious Romans because it was believed to be an antidote to poisoning by hemlock or deadly nightshade.

THE FIBROUS STALKS OF THE nettle are stronger than flax and can be spun into a yarn to make a **very durable material**. Over the years nettles have also been used to make clothes, fishing lines, rope and bedlinen. They also make fine **window cleaners**. Take a bunch, wearing gloves of course, and plunge it into a bucket of water with a tablespoon or two of vinegar. Rub over the windows and then dry with a cloth.

Five sleep-inducing teas

- Bergamot flowers
- Camomile flowers
- Hop flowers
- Lemon balm
- Lemon verbena

Give yourself a refreshingly fragrant bath by filling a small muslin bag with dried camomile flowers or lavender or rosemary. Hang the bag around the hot tap so that the water runs through it, or soak it in boiling water for 10 minutes and then drop it into the bath. Lavender is mildly sedative and good to use if you want to feel relaxed.

CAMOMILE FLOWERS SHOULD be dried when the flowers first open, usually in July. Tie the stems and hang them upside down.

IF YOU WANT TO LIGHTEN your hair, take half a cup of **camomile flowers**, fill it up with boiling water and leave it to infuse for an hour or so before rinsing your hair with it. A similar infusion of rosemary leaves adds lustre to darker hair.

MUGWORT, ALSO KNOWN AS midge plant, works as a good insect repellent if you make an infusion from the leaves and rub it into the skin. Camomile flowers will also do the job.

Camomile lawns were popular in Tudor times. Drake's famous game of bowls before he set sail to defeat the Armada was played on a camomile lawn.

FENNEL IS A STRIKING PLANT TO have in your garden border, but it's the **neighbour from hell** for a number of other plants, especially its umbelliferous cousins coriander and dill. Fennel, though, has a more harmonious relationship with babies. Herbalists use an infusion of the seeds to help the symptoms of colic, and you can also rub fennel and camomile oil into the baby's skin for the same purpose. Fennel tea helps to stimulate the flow of breast milk and some mothers add it to bottles of milk to aid digestion. This pungent herb is also an effective treatment for chest congestion and is a common ingredient in cough remedies for adults and children.

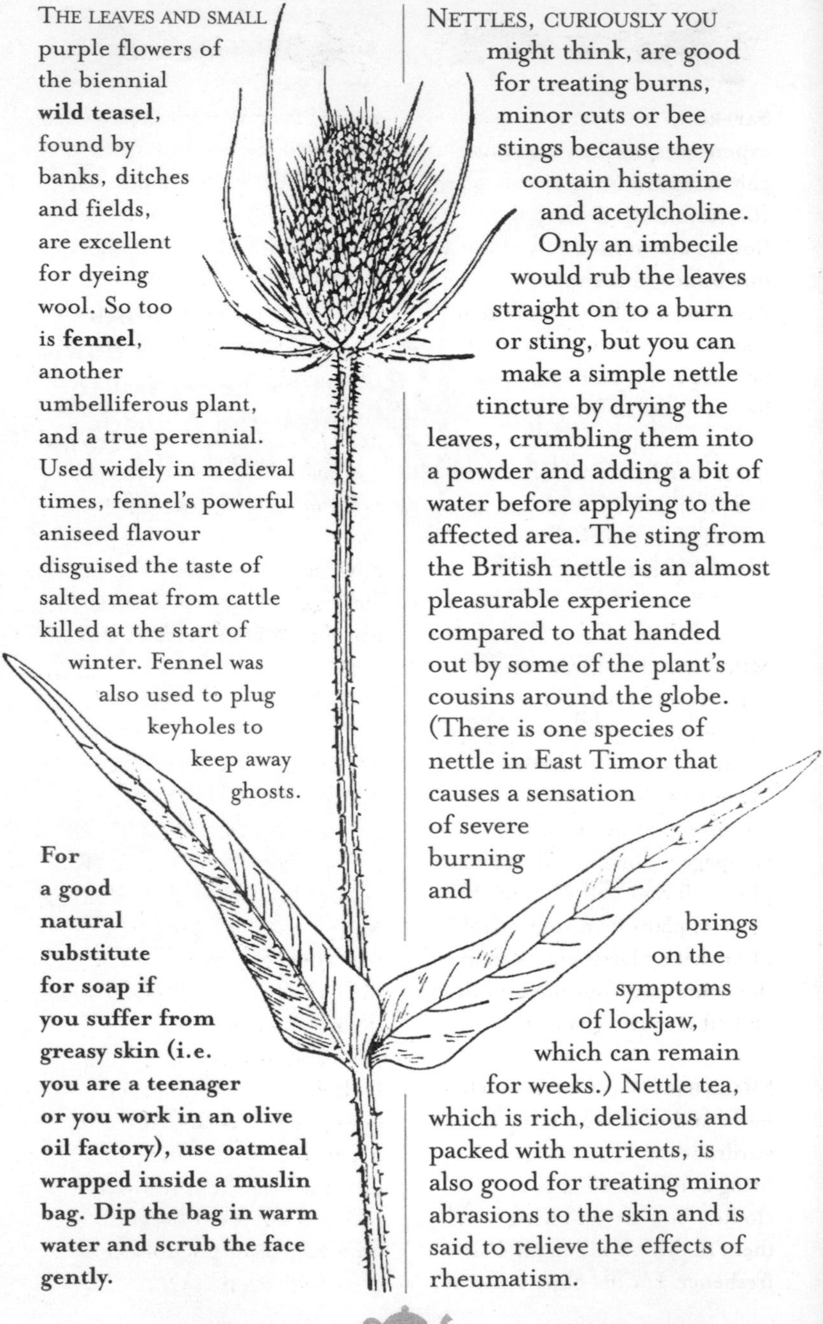

THE LEAVES AND SMALL purple flowers of the biennial **wild teasel**, found by banks, ditches and fields, are excellent for dyeing wool. So too is **fennel**, another umbelliferous plant, and a true perennial. Used widely in medieval times, fennel's powerful aniseed flavour disguised the taste of salted meat from cattle killed at the start of winter. Fennel was also used to plug keyholes to keep away ghosts.

For a good natural substitute for soap if you suffer from greasy skin (i.e. you are a teenager or you work in an olive oil factory), use oatmeal wrapped inside a muslin bag. Dip the bag in warm water and scrub the face gently.

NETTLES, CURIOUSLY YOU might think, are good for treating burns, minor cuts or bee stings because they contain histamine and acetylcholine. Only an imbecile would rub the leaves straight on to a burn or sting, but you can make a simple nettle tincture by drying the leaves, crumbling them into a powder and adding a bit of water before applying to the affected area. The sting from the British nettle is an almost pleasurable experience compared to that handed out by some of the plant's cousins around the globe. (There is one species of nettle in East Timor that causes a sensation of severe burning and brings on the symptoms of lockjaw, which can remain for weeks.) Nettle tea, which is rich, delicious and packed with nutrients, is also good for treating minor abrasions to the skin and is said to relieve the effects of rheumatism.

Saffron, the world's most expensive spice by weight, is cultivated from the crocus plant (*Crocus sativus*). Over 4,000 flowers are needed to yield just one ounce of the spice. The Essex town of Saffron Walden takes its name from the spice because it was cultivated as a harvest crop in the region.

If they would drink nettles in March
And eat mugwort in May
So many young maidens
Wouldn't go to the clay.

Old saying

Mint, which spreads by runners and seed, can grow 20 feet underground and pop up in an unexpected corner of your garden. To prevent the spread of mint, which will **rampage** through your vegetable plot or flower bed if given the chance, plant it in some kind of bucket or large pot. Remove the container's bottom and then embed it in the ground.

Small bags filled with dried herbs and/or flowers hung in wardrobes and cupboards will bring a fresh smell to your clothes. You can also hang them in the loo as a natural air freshener. For the bags, use thin

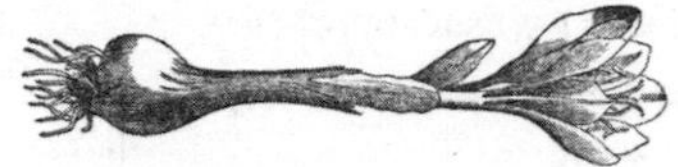

natural materials that breathe well such as cotton, muslin or linen and tie them up with a piece of cotton or, if you're feeling fancy, some ribbon. The old-fashioned **lavender bag** traditionally included lavender flowers, dried thyme and mint and a dozen or so ground cloves.

Saint John's Wort with his flowers and seed boiled and drunken, provoketh urine, and is right good against the stone in the bladder . . . The leaves stamped are good to be laid upon burnings, scaldings, and all wounds; and also for rotten and filthy ulcers.

John Gerard
The Herball, 1597

There are two good reasons for **spreading your herbs** around the garden as well as, or instead of, lumping them all together in a herb garden. Many herbs have properties that encourage growth in other plants and they can also protect them from harmful insects, pests and diseases. Examples include sage with cabbages, borage with strawberries, and marigolds, which exude an evil smell repellent to insects, with any number of vegetables, especially peas, potatoes and tomatoes (see p. 44).

YARROW, THE WITCH'S HERB used in spells and incantations, is good for stemming bleeding and healing wounds. It is said to have been the plant used by Achilles to treat the wounds of his soldiers – hence the proper name *Achillea millefolium* given to the plant in his honour. It was used by soldiers as recently as the First World War and its various nicknames – staunchweed, soldier's wort, bloodwort – celebrate its healing properties. Many older gardeners keep some in the shed in case they cut or graze themselves. Yarrow dries well and is good for winter flower arrangements. If you plan to grow it, bear in mind that it's rather smelly and also quite invasive and needs to be controlled. Yarrow's astringent properties make it a good **face wash** for those who suffer from oily skin. Put some dried flowers in a large cup of boiling water, let it cool before straining it and then dab it on the skin.

Mint will grow in most conditions but it prefers a slightly damp, semi-shaded area. It will struggle if grown in a container in full sunshine. Most herbs love a site in the glare of the sun but others that enjoy a moister, shadier site are bergamot and angelica.

MINT'S MEDICINAL AND culinary properties have been championed from ancient times to the present day. The menthol in **peppermint soothes** the lining of the digestive tract and stimulates the production of bile, which is essential for good digestion. A hot cup of fresh peppermint tea is an excellent way to settle your stomach after a big meal.

THE YELLOW-FLOWERED **St John's wort** is a common feature of British gardens and recently has been used to treat mild to moderate depression and anxiety disorders. In Germany it is prescribed far more frequently than pharmaceutical antidepressants.

In order to live off a garden you practically have to live in it.

FRANK MCKINNEY HUBBARD (1868–1930)

To make a strong, sweet-smelling, old-fashioned **pot pourri**, pick a combination of bergamot, geranium, roses, rosemary, hyssop and lavender – or any other strong-smelling flowers that dry well. Adding the leaves of pungent herbs such as lemon verbena, lemon balm, mint, thyme and marjoram will boost the fragrance. Pick in the morning after the dew has evaporated but before it gets too hot, remove the stems and leave the flowers and leaves to dry. When the flowers are completely dry, break up the petals and put them and the leaves into a jar with a sprinkling of sweeter spices like cinnamon and nutmeg, together with a dash of rock salt. Leave for about a fortnight, giving it the odd shake before transferring into bowls.

In days gone by, **pillows** filled with dried herbs were a common feature in the British bedroom. They were used as a cure for insomnia as well as a means of filling the room with a pleasant, relaxing smell. Lavender, marjoram, rosemary and camomile were all common ingredients, but pillows filled only with dried hops were also popular. **Hops** contain the natural tranquillizer lupulin (the proper name for the hop is *Humulus lupulus*) and it is said that **George III** cured his insomnia during his long illness by taking a hop pillow to bed.

Kent, sir – everybody knows Kent – apples, cherries, hops and women.

Charles Dickens
Pickwick Papers, 1837

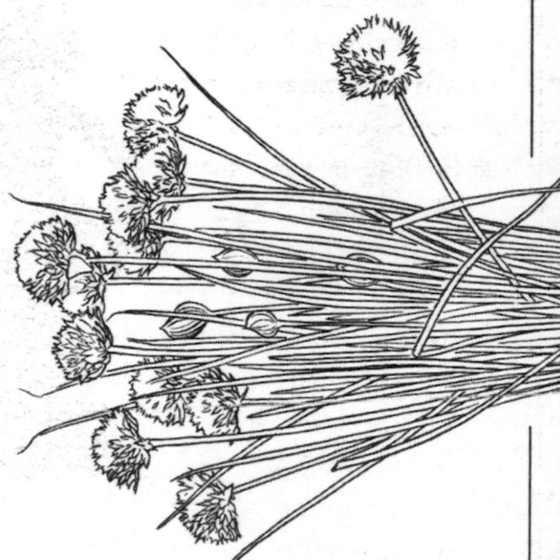

Chives are the smallest member of the onion family.

Lavender, *Lavandula*, comes from the Latin *lavare*, 'to wash', because its oils were often used in cleaning clothes.

There's rosemary, that's for remembrance; pray, love, remember: and there is pansies, that's for thoughts.

William Shakespeare
Hamlet, 1601

The cold winds of March can kill herbs such as thyme and blister and discolour the leaves of rosemary and bay. Ideally, a herb garden should face south-west and be sheltered from harsh winds from the north and east, especially if you live in East Anglia.

HOWARD CARTER AND HIS archaeological team found many seeds and plants stored in the tomb of **Tutankhamun** in the Valley of the Kings, including garlic, cumin, coriander, watermelon, wheat, barley, lentils, flax, fenugreek, chickpeas, olive oil, almonds and dates.

ROCKET IS BACK IN FASHION after centuries of neglect. After arriving in Britain with the Romans, it was grown widely until the eighteenth century and valued because it is so easy to grow, is ready to eat just eight weeks after sowing and grows back after cutting. The word is thought to derive from the Latin word *eruca*, meaning 'caterpillar', because of its slightly furry stems.

NETTLES ARE GOOD FOR YOUR garden. Like Russian comfrey, they make an excellent mulch or liquid feed, and because they are so rich in nitrogen the feed is especially effective on plants that demand strong leaf growth. They store many vital nutrients from the soil, including nitrogen, iron, silica, protein and phosphate. Turn them into a fertilizer and you'll be giving your plants a feast. It's very easy to make: just fill a container with nettles and rainwater and leave for about four weeks. Dilute it by about 10 parts to one and pour it on to the roots of the plants.

FENNEL SALAD, WHICH IS VERY popular in Italy, makes an excellent and delicious aid for digestion at the end of a filling meal. Take a few fennel bulbs, cut off their bottoms, slice thinly and dress with olive oil, salt and pepper. This salad goes well with slices of orange that have been peeled, de-pithed and de-pipped.

THE SEEDS OF **DILL** AND **CARAWAY** were once a popular form of chewing gum.

Herbs are the friend of the physician and the pride of cooks.

CHARLEMAGNE (747–814)

IF YOU PLAN TO GROW HERBS on your **windowsill** indoors, remember that the humidity is often low in centrally heated houses. You can mitigate the dehydrating effects by squeezing as many plants as possible on to the sill, as the plants and their compost both exude moisture.

TO MAKE ONE AND A HALF pounds of **lavender oil**, you'll need to harvest almost a quarter of a ton of flowers.

THE ROOTS AND LEAVES of the **saponaria** plant, colloquially known as soapwort and still a common sight throughout British gardens and countryside, were for many centuries used as a **soap substitute**. They produce a lather when rubbed in water.

He who has sage in his garden will not die.

ARABIC PROVERB

AS ANY PASSING TOOTHPASTE manufacturer will be only too happy to tell you, mint leaves make a first-rate **breath freshener** and are that much more effective if chewed with a piece of aniseed or cinnamon. Mint also helps relieve some of the pain caused by the mouth condition gingivitis.

TO MAKE **FLAVOURED VINEGARS** with herbs, use wine or cider vinegar for best results. Pick the herbs in the morning once the dew has gone and the sun has had the chance to steal the oils from the leaves. As with flavoured oils, simply fill a jar with herbs, cover with vinegar and leave for a few weeks, or follow the heating method outlined on p. 93. After straining the vinegar into a bottle, pop in a fresh sprig because it looks good and will remind you of its flavour if you have a number of different ones in your cupboard.

10 common herbs suitable for making flavoured vinegars

- Basil
- Chervil
- Chives
- Dill
- Garlic
- Marjoram
- Mint
- Rosemary
- Sage
- Tarragon

THE CULINARY QUALITIES OF **common sorrel**, like the common nettle, were once widely recognized by Britons but have been largely forgotten in recent years. Sorrel leaves look very like the dock leaves to which they are related. Both grow abundantly throughout Britain. Sorrel is often found on wasteland and in meadows, but you may well have some growing in your back garden without realizing it. Sorrel, which is high in vitamin C, has a slightly tart, lemony flavour and adds extra piquancy to salads. It also works well as a basic green vegetable when steamed or very lightly boiled. It was once a common ingredient in sauces to accompany fish and is still widely used in France in various dishes including soups and salads. Those who know their sorrel well, however, say that it is best enjoyed in an omelette, fried in butter for five minutes or so before the egg mixture is added. Why sorrel is also called donkey's oats is anyone's guess because donkeys don't actually like it.

MINT JULEP COCKTAIL

For a refreshing summer sundowner after a hard day's work, you could do worse than pour yourself one of these ✿ Put about five mint leaves in a long glass with a tablespoon of sugar (the finer the better) and a tablespoon of water, and crush it all up to release the flavour of the mint ✿ Add a decent slug of bourbon (Jack Daniel's, Jim Beam, etc.), fill the glass with crushed ice, garnish with a sprig of mint and drink through a straw.

I plant rosemary all over the garden, so pleasant is it to know that at every few steps one may draw the kindly branchlets through one's hand and have the enjoyment of their incomparable incense; and I grow it against walls, so that the sun may draw out its inexhaustible sweetness to greet me as I pass.

Gertrude Jekyll (1843–1932)

MINT SAUCE

Take a couple of handfuls of leaves and chop them up, either in a processor or by putting them into a mug and going at them with a pair of scissors ✿ Bring some vinegar to a simmer in a pan and add some sugar if you want a sweeter mixture, or some water if you want to reduce the sharpness ✿ Then add the chopped leaves, allow to cool, pour into jars or bottles and store. If you keep it in the fridge the sauce can last up to eight months.

As for the garden of mint, the very smell of it alone recovers and refreshes our spirits, as the taste stirs up our appetite for meat.

Pliny the Elder (23–79)

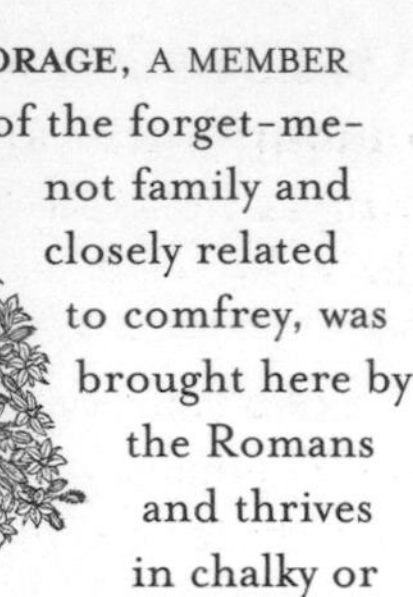

Borage, a member of the forget-me-not family and closely related to comfrey, was brought here by the Romans and thrives in chalky or sandy soil. The leaves, which have the same mild taste as cucumbers, can be used in Pimm's or summer cordials, cooked like sorrel for greens, or used in salads or as the garnish on a potato salad (the blue flowers can also be added). It is often grown as a companion for strawberries. The word is thought to come from the Arabic, meaning 'father of sweat', because doctors used it to encourage perspiration.

To **make your own mustard**, crush up your seeds (black, white or a combination of the two) and add the powder to a small amount of cold water and the same amount of flour. (Hot water kills the enzymes and produces a nasty bitter flavour.) You can add what you want from here on, but for tanginess mix in some wine vinegar and/or horseradish. For sweetness add some honey, brown sugar or fruit juice and for different tastes add some chopped-up or blended herbs, or some spices.

Mint is an excellent **insect repellent** and growing some in a window box will deter flies from coming into the house. Rub some mint leaves into your face and neck on a hot day to stop them buzzing round your head when you are in the garden. Boil up some camomile flowers and apply the cooled liquid to your exposed skin for similar protection. Sage, meanwhile, has the same repellent effect on midges. If you are in the garden, tie some round your neck or stick it behind your ears, but try to remember to remove it if you are going out for the evening.

The garden is the poor man's apothecary.

German proverb

Wort is an old Saxon word meaning 'herb'. Dragon's wort is another name for tarragon. French tarragon, a key ingredient of Béarnaise and tartare sauces, has a much finer flavour than Russian, which can often be virtually tasteless. After a few years the flavour of French tarragon tends to diminish. The herb has nothing to do with France or Russia: it was brought to Spain by the Moors.

As for rosemary, I let it run all over my garden walls, not only because my bees love it, but because it is the herb sacred to remembrance and therefore, friendship.

St Thomas More (1478–1535)

A basil leaf placed on a **mouth ulcer** and left as long as possible will ease the ache.

Over **80 per cent** of the world's population still relies on herbs for everyday medical treatment.

Herbs suitable for container growing

Basil (better indoors)
Chives (will put up with a bit of shade)
Coriander (doesn't like wind)
Dill (pretty tolerant but likes the sun)
French tarragon (likes a good deal of shade)
Garlic (plant in autumn so it gets the cold it needs to grow)
Lemon balm (very unfussy)
Mint (prefers only a little sun)
Oregano/marjoram (lots of sun)
Parsley (difficult and slow to germinate)
Rosemary (lots of sun)
Sage (perfectly happy in hot, dry weather)
Sweet Florence fennel (smaller than common fennel)
Thyme (lots of sun)

Herb butter is great with almost anything. Classic combinations include rosemary or sage for chicken, fennel for fish, basil for fish or vegetables, tarragon for fish or steak and mint for lamb, potatoes or peas. It's also very simple to prepare. Soften up the amount of butter you want in a warm bowl with a fork, add a squeeze of lemon and two or three tablespoons of the chopped herbs. Shape it into a roll and wrap it in clingfilm or foil, and keep it in the fridge to be sliced off as and when you need it.

Plant me a garden to heal the body
Betony, yarrow and daisies to mend
Sage for the blood and comfrey for bones
Foxglove and hyssop the sick to tend.

Chorus
Tansy, rosemary, rue and thyme
Bring back the lover who once was mine
I will give him the sweet basil tree
Then he will always belong to me.

Plant me a garden to heal the heart,
Balm for joy, and the sweet violet
Cowslips, pansies and chamomile
To ease the pain I want to forget.

Plant me a garden to heal the soul,
A garden of peace and tranquillity,
Soothed with the scent of lavender
And the heavenly blue of chicory.

Elizabethan herb song

Greek athletes ate sweet fennel to build up their stamina and give them courage. The Greek word for fennel is *marathon*. The plant grew wild around the small town near Athens that gives its name to the race we know today. Roman soldiers used fennel for the same reason and its medicinal properties were so highly regarded that Pliny the Elder, a natural philosopher, recommended it in over 20 remedies. Frankish emperor Charlemagne insisted that fennel was grown in all his imperial gardens.

Salome was thought to have hidden **the head of St John the Baptist** in a bed of basil, while in John Keats's poem 'Isabella' (opposite), a young woman conceals the head of her murdered lover in a pot in which she plants the 'king of herbs'.

LAVENDER LEMONADE IS A deeply refreshing, old-fashioned drink for a hot summer's day. Bring about 700ml of water to the boil with a mug of sugar (you can add more sugar if you like it a bit sweeter), add the flowers from about 15 lavender stems and turn off the heat. When the mixture has cooled add another 700ml of cold water and squeeze in the juice of about three lemons. Strain out the lavender flowers and serve with ice and a fresh lavender flower.

VALERIAN, DERIVED FROM THE Latin word *valere*, meaning to be strong or brave, grows wild in woodlands, damp meadows and along riverbanks throughout Europe and is cultivated in Britain, Holland and Belgium. Its root is used to relieve insomnia and nervous tension. Until the 1940s it was administered as a tranquillizer and it was used by the British to treat shell shock during the First World War and to treat civilians traumatized by air raids in the Second.

Samuel Pepys liked nettle porridge for breakfast.

BASIL MAYONNAISE IS A fantastic relish for hamburgers and sandwiches and in the United States many cooks like to baste their chicken with it before roasting. You make it simply by whizzing up a handful of basil leaves in a blender with some plain mayonnaise and crushed garlic.

THE **MINT FAMILY** INCLUDES thyme, sage, lavender, marjoram, oregano, rosemary, hyssop, basil, catnip, betony and horehound.

And she forgot the stars, the moon, and sun,
And she forgot the blue above the trees,
And she forgot the dells where waters run,
And she forgot the chilly autumn breeze;
She had no knowledge when the day was done,
And the new morn she saw not: but in peace
Hung over her sweet Basil evermore,
And moistened it with tears unto the core.

JOHN KEATS
'Isabella; or, The Pot of Basil', 1820

How I would love to be transported into a scented Elizabethan garden with herbs and honeysuckles, a knot garden and roses clambering over a simple arbour.

ROSEMARY VEREY
celebrated British gardener
(1919–2001)

TO MAKE **LAVENDER SOAP** GRATE some good-quality, plain, natural soap and put in a bowl with a cup of boiling water. Place the bowl in a pan of hot water and stir until smooth. Crush about half a cup of dried flowers and, having removed the bowl from the pan, add them to the soap and put in a few drops of lavender or almond oil. You can mould the soap with your hands and leave the bars to dry on wax paper or a metal tray.

Four spinach alternatives in your garden

Angelica
Borage
Nettles
Sorrel

Foxglove is an important medicinal herb and today the leaves are used in the manufacturing of a drug (digitalis) for the strengthening and normalization of the heartbeat. One plant will produce thousands of seeds and in the spring there will be dozens of seedlings emerging around the old plant. In spite of its medicinal properties, foxglove is highly poisonous to humans and animals but it rarely causes death because it tastes so foul that there is little temptation to carry on chewing it.

Hops were introduced to Britain in the early 1500s by Flemish weavers settling in Kent to work in England's booming wool industry. Henry VIII banned the use of hops in brewing, but in 1552 his son Edward VI oversaw legislation to permit their use. Until then, ale was made with malted barley, spices, herbs and even tree bark, and would have tasted far sweeter than the beers we drink today.

An old-fashioned and natural way of **strengthening the hair** and adding shine is to make an infusion with some nettle tops. Place them in a large cup of boiling water, cover and leave for a few hours. Take out the tops and rinse your hair with the cool liquid after shampooing.

Alaric, King of the Visigoths, demanded 3,000 pounds of black pepper as part of the ransom for Rome during the first siege of the city. Rome eventually fell on 24 August 410 after the third siege.

A bed of **watercress**, an aquatic perennial, can yield up to 10 harvests a year provided the water is kept clean.

Five common herb/flower oils with reputedly antidepressant properties

Bergamot
Camomile
Clary sage
Geranium
Rose

Flowers

Don't send me flowers when I'm dead

People from a planet without flowers would think we must be mad with joy the whole time to have such things about us.

Iris Murdoch (1919–99)

The **daisy** got its name because early Anglo-Saxons thought that the yellow centre resembled the sun. They knew it as *daeges eage* and by the late Middle Ages it was commonly known as the day's eye until the name was transmuted into the word we recognize today.

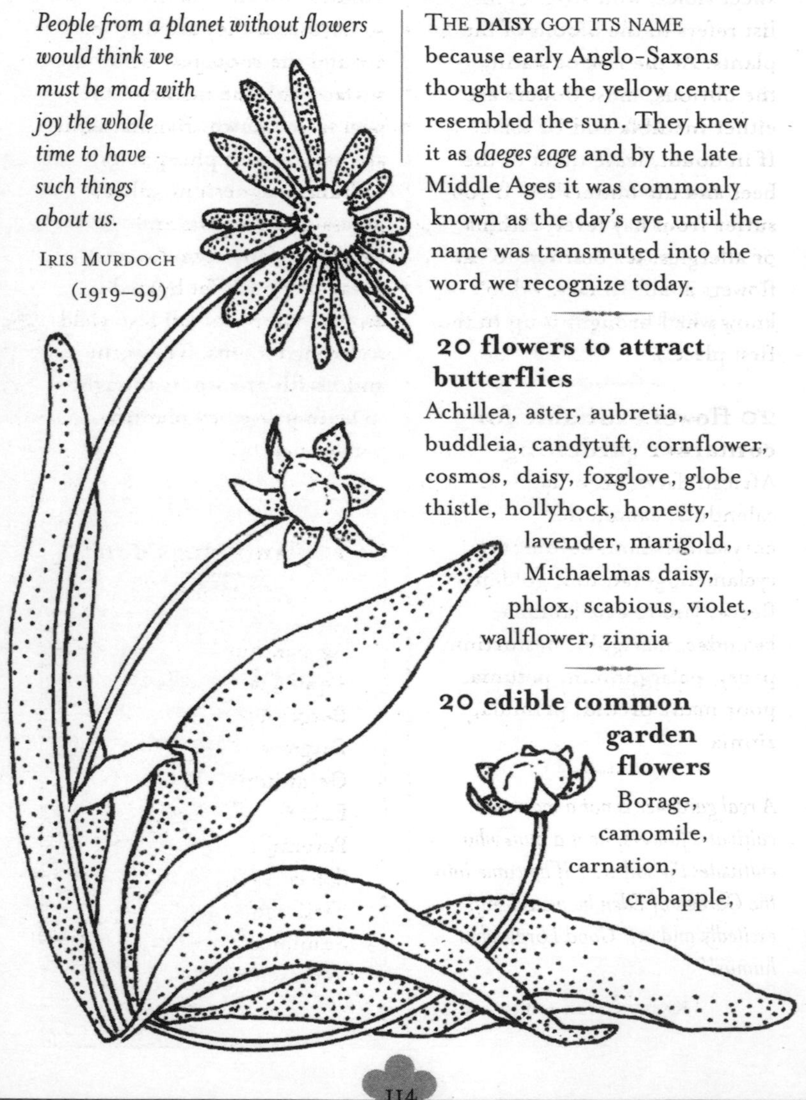

20 flowers to attract butterflies

Achillea, aster, aubretia, buddleia, candytuft, cornflower, cosmos, daisy, foxglove, globe thistle, hollyhock, honesty, lavender, marigold, Michaelmas daisy, phlox, scabious, violet, wallflower, zinnia

20 edible common garden flowers

Borage, camomile, carnation, crabapple,

cornflower, daisy, dandelion, elder, hawthorn, hibiscus, honeysuckle (but definitely not the poisonous berries that they become!), hyssop, lavender, marigold, nasturtium, pansy, primrose, sunflower, sweet violet, wild rose. (This list refers to the bloom of the plants. At the risk of stating the obvious, most flowers are either inedible and/or toxic. If in doubt, leave them to the bees and the butterflies. If you suffer from hay fever, asthma or allergies, it's best not to eat flowers at all. In fact, I don't know why I brought it up in the first place.)

20 flowers suitable for container gardening

African daisy, begonia, calendula, camomile, chrysanthemum, columbine, cyclamen, geranium, golden fleece, lamb's ear, lantana, lavender, marigold, nasturtium, pansy, pelargonium, petunia, poor man's orchid, primula, zinnia

A real gardener is not a man who cultivates flowers; he is a man who cultivates the soil . . . If he came into the Garden of Eden he would sniff excitedly and say 'Good Lord, what humus!'

KAREL ČAPEK (1890–1938)
The Gardener's Year

PRINCE CHARLES, AKA THE Duke of Cornwall, is paid one daffodil a year as rent for unattended lands on the Scilly Isles.

OLD BANANA SKINS CAN DO wonders for the quality of your roses if they are dug in around the roots just below the surface with the inside of the peel facing down. Banana skins are packed with phosphates, sodium, magnesium, silica, potassium, sulphur and calcium. Many rose-lovers swear that meat fat buried around the roots will also yield stunning blooms. The scent and health of roses is thought to be improved by planting parsley nearby.

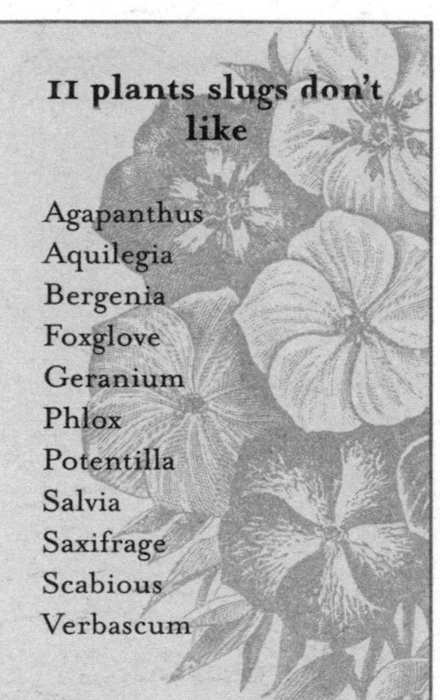

11 plants slugs don't like

Agapanthus
Aquilegia
Bergenia
Foxglove
Geranium
Phlox
Potentilla
Salvia
Saxifrage
Scabious
Verbascum

Daffodils are to spring what Roses, Irises and Lilies are to summer, what Sunflowers and Chrysanthemums are to autumn and Hellebores and Aconite are to winter.

WILLIAM ROBINSON
(1838–1935)
The English Flower Garden

ORNITHOGALUM, OR STAR OF Bethlehem, is sometimes known as **florist's nightmare** because the flowers can last for as long as a month. The word *ornithogalum* means 'bird's milk' in Greek – an expression used to describe something unbelievable. In England these attractive, fragrant white flowers with long stems are also called chincherinchee, which is a corruption of the tongue-twisting name *tjenkenrientjee* by which they are known in South Africa.

Earth laughs in flowers.

RALPH WALDO EMERSON
(1803–82)

THE **LARGEST INDIVIDUAL flower** in the world is the *Rafflesia arnoldii*, which can grow up to a metre in diameter and weigh as much as seven kilos. Each fully grown petal measures roughly half a metre. It grows on the Indonesian island of Sumatra as well as in Malaysia and Borneo. It is named after two Britons, **Sir Stamford Raffles**, the founder of the British colony of Singapore, and his friend Dr Joseph Arnold, who discovered the giant flower in 1818. The plant emits a stench of rotting flesh in order to attract insects for pollination.

At present, I am mainly observing the physical motion of mountains, water, trees and flowers. One is everywhere reminded of similar movements in the human body, of similar impulses of joy and suffering in plants.

EGON SCHIELE (1890–1918)

IF YOU HAVE A **SMALL GARDEN**, place your bright, colourful flowers or larger-leaved plants nearer the house rather than at the end of your plot of land because the eyes are immediately drawn to bright colours and bigger objects. If you put them at the end of the garden they will have the effect of making it seem smaller than it is.

Each mature sunflower produces 40 per cent of its weight in oil.

On the sunflower: . . . *bears abundance of oily seed, much liked by poultry of every sort . . .*
Fit for nothing but very extensive shrubberies. When seen from a distance the sight may endure it.

William Cobbett (1763–1835)

The **sunflower** is a native plant of the Americas, where the Indians used its seed as an important source of food. The Incas of Peru were sun worshippers and used it in religious ceremonies. The sunflower leans towards the sun in a process known as heliotropism.

Sunflower stems were once used to **fill lifejackets** before the invention of synthetic materials.

Shed No Tear! O shed no tear!
The flowers will bloom another year.
Weep no more! O weep no more
Young buds sleep in the root's white core.

John Keats (1795–1821)
'Faery Song'

Umbelliferous plants, whose flowers are arranged **like an umbrella** or the rose of a watering can, are conspicuous and attractive to many beneficial insects. (The word umbellifer derives from the Latin for parasol.) The large flower heads provide a comfortable, efficient landing pad for the insects, allowing easy access to pollen. The umbellifer family boasts around 3,000 species including angelica, anise, carrot (wild), chervil, coriander, cow parsley, cumin, dill and fennel.

10 flowers providing lots of seeds for birds

Angelica	Fennel	Snapdragon	Zinnia
Cornflower	Lavender	Sunflower	
Cosmos	Scabious	Teasel	

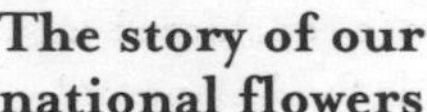

The story of our national flowers

The story goes that when Vikings approached to attack the sleeping Scottish army under cover of darkness, one of their number stepped on a **thistle** and let out a very un-Viking-like screech, which woke up the Scots, who promptly got out of bed and routed the invaders. In recent years the **daffodil** has slowly come to rival the leek as the national emblem of Wales. No one knows exactly why this should be so, but one theory is that the Welsh have become fed up with trying to slot long vegetables into their buttonholes on St David's Day. The fact that there is something faintly comic about the leek may also play a part. Perhaps it lacks the grandeur and elegance of the thistle, the rose or the lily, while the daffodil is unquestionably beautiful. How the leek ever came to be adopted as the symbol of the Welsh nation remains a mystery. Some attribute it to the country's St David who, as a monk, was said to have lived on leeks and bread, while others claim it harks back to the Battle of **Agincourt**, where the leek was worn to identify the Welsh archers in Henry V's army. Another explanation is that it identified the followers of Welshman Henry Tudor, whose coat of arms consisted of green and white colours.

Tudor became Henry VII after defeating Richard III at the Battle of Bosworth Field in 1485, thus bringing to an end the 30-year Wars of the Roses. If the previous suggestion is true, then Henry VIII's father can congratulate himself on having created two national symbols because it was following his succession to the throne that the rose became the national symbol of England. Henry Tudor fought in the name of the House of Lancaster (symbol: red rose) against the followers of the House of York (white rose), but when he married Elizabeth of York he united the two families and created a national symbol in the **Tudor rose** – a red rose with a white centre.

THE FORMAL NAME FOR THE Scotch thistle is *Onopordum acanthium*, but it used to be known more colloquially as **farting donkey**, from the Greek *ono* (donkey) and *porde* (fart).

DAFFODIL BULBS CAN last for up to 80 years.

IN DAYS GONE BY PEOPLE WHO kept chickens treated the **daffodil** with suspicion as they believed it would stop their hens, ducks and geese from laying eggs.

THE NETHERLANDS WERE convulsed by a **mania for tulips** in the 1630s and staggering sums of money were paid for their bulbs. Some are recorded as having been sold for over 5,000 guilders – the equivalent of 20 years' wages for a London bus driver today.

DAFFODIL BULBS CONTAIN A substance called galanthine, which scientists have developed for use in the treatment of Alzheimer's. It has also been used as an antispasmodic to treat hysteria and epilepsy.

THE **STAR OF BETHLEHEM flower** is also known, *inter alia*, as the eleven o'clock lady, wake-at-noon, Jack-go-to-bed-at-noon and sleepy dick because it does not flower until about lunchtime and then closes up for the day at about three (nice work if you can get it). The plant is related to the onions and chives and the bulbs were once considered to be a nutritious food plant and delicacy in the Middle East, where they are known today as **dove's dung**.

It is a greater act of faith to plant a bulb than to plant a tree.

CLARE LEIGHTON (1901–89)

LAD'S LOVE (*Artemisia abrotanum*), known also as maiden's ruin or southernwood, was once a symbol of love and fidelity. Traditionally, country boys used to give it to their sweethearts after church, the plant's powerful aroma helping to disguise their own, less attractive odours. It became a common feature in cottage gardens after its introduction to Britain from southern Europe in the mid-sixteenth century and it was thought to bring good luck to those living in the house. It was also used as a moth repellent when dried, hence the French name for it, *garde-robe* or 'wardrobe'.

PRIMROSES WERE ONCE MADE into puddings by frying the flowers in butter and sugar. Prime Minister **Disraeli** was said to have enjoyed the dish for his breakfast.

IT IS STRANGE THAT **ORCHIDS** ARE still considered rare when it has been known for some time that there are thousands of species, possibly as many as **30,000**, and even more hybrids, making the family the largest of all flowering plants. The variety of orchids is equally astonishing and they can be found in the Arctic as well as around the Equator, some with flowers the size of a pencil tip, others larger than a football.

THE PETALS OF THE **ORCHID** *Trichoceros parviflorus* resemble a particular female fly so closely that the male of the species will try to mate with it – and thus pollinate the orchid.

THE PODS OF SOME ORCHIDS hold as many as three million seeds.

IN 1913 A GROUP OF **Suffragettes** invaded Kew Gardens, attacked the Orchid House and burned down the Tea Pavilion. Suffragette co-founder Emily Pankhurst commented on the ongoing campaign of violence: *We are not destroying Orchid Houses, breaking windows, cutting telegraph wires, injuring golf greens, in order to win the approval of the people who were attacked. If the general public were pleased with what we are doing, that would be a proof that our warfare is ineffective. We don't intend that you should be pleased.*

BLACK FLOWERS WERE CHERISHED by Victorians and Edwardians, especially by the Art Nouveau movement, but in truth there is no such thing. Black tulips are very dark purple and black roses are very dark red.

When a Frenchman reads of the Garden of Eden, I do not doubt but he concludes it was something approaching to that of Versailles, with clipt hedges, berceaus and trellis work.

HORACE WALPOLE
Essay on Modern Gardening, 1785

THE BOTANICAL NAME FOR THE **snapdragon** is *Antirrhinum*, which comes from the Greek meaning 'resembling a nose'. These popular flowers have a number of other colourful nicknames including calves' snouts, lion's lips and toad's mouth.

True that a plant may not think; neither will the profoundest of men ever put forth a flower.

DONALD CULROSS PEATTIE
(1896–1964)
botanist and author

Origins of the names of some flowering plants (and, er, one tree)

Acer (maple). Acer is Latin for sharp. Romans used the tree to make arrows.

Aquilegia. From Latin *aquila* for eagle. So called because of its wing-shaped petals.

Buddleia. Named after the seventeenth-century English botanist Adam Buddle.

Campanula. From Latin *campana* for bell. Flowers are bell-shaped.

Forsythia. After the eighteenth-century Scottish gardener and writer William Forsyth.

Fuchsia. After the sixteenth-century German botanist and herbalist Leonard Fuchs.

Orchid. From Greek *orchis* for testicle. Bulbs of many species are shaped like testicles.

Philodendron. From Greek *phileo* for I love, and *dendron* for tree.

Phlox. From Greek word for flame. Brightly coloured flowers.

Raphanus (radish). From Greek *ra* for quick, and *phainoma*, to appear. Fast-growing vegetable.

Sedum. From Latin *sedere*, to sit. Low-lying plant.

Large or small it [the garden] *should look both orderly and rich. It should be well fenced from the outer world. It should by no means imitate the wilfulness or the wildness of Nature, but should look like a thing never seen except near a house.*

WILLIAM MORRIS (1834–96)
Hopes and Fears for Art

DANDELION LEAVES MAKE FOR A good salad – and they're also a delicious substitute for lettuce in a bacon sandwich. Dandelion leaves that have not been blanched are only truly tasty when eaten young. Blanching the older leaves removes the bitterness: tie the leaves together where they are growing and cover them with a pot for about a week. For a good greens dish, steam some leaves and eat with butter and lemon juice.

Dried nasturtium seeds were ground into a powder during the Second World War as a replacement for pepper.

IN BERKSHIRE AND Worcestershire, dandelion flowers are used to make **dandelion wine**, which tastes a little like sherry and has been used for centuries as a highly effective tonic. There are dozens of different recipes for dandelion wine, but they are all variations on a central theme. The following is a simple, traditional one. They say you should pick the flowers when the sun is at its highest because then the flowers will be fully open. Pick two litres or more of flowers with all the greenery removed. (It might be a good idea to get your children to do this bit, freeing you up to do more important things like read the newspaper or have a snooze.) Put the flowers in a big pan and pour over four litres of boiling water, cover with a cloth and leave for two days, but no longer. Add the peel and juice but not the pith of about four oranges, together with 1.2 kilos of sugar, and bring to a boil for about 20 minutes. Add 15g dried yeast and let the mixture ferment in a container for about a week, then bottle. The longer you leave it the better, but it will be particularly good if you leave it for six to 12 months.

A SINGLE **DANDELION** FLOWER turns to about 180 seeds, but an established three-year-old plant produces as many as 5,000.

If dandelions were hard to grow, they would be most welcome on any lawn.

ANDREW MASON (N.D.)

The use of dandelion as a caffeine-free **substitute for coffee** has been increasing significantly in recent years. (The Germans have even come up with an instant version of it.) It tastes remarkably similar to the original, but it is expensive to buy. Homemade dandelion coffee is easy: all you need is a plentiful supply of dandelions from your garden or a nearby green or playing field, preferably picked in the autumn when the roots are at their fullest. Dig up as many roots as you can find (carefully avoiding the hairy-stemmed imposters) and simply wash then dry them in the sun, roast in the oven, then grind them and use as you would coffee, adding milk, sugar or honey to taste.

Earwigs love eating flowers and they're particularly fond of dahlias, clematis and chrysanthemums. Like slugs and snails, they generally come out only after dark and are difficult to catch *in flagrante delicto*. One way to catch them is to take a flowerpot, stuff it with straw and suspend the pot upside down on a small bamboo pole among the flowers. In the morning, you can murder them in their sleep.

All garden plants derive from wild versions which have adapted themselves to particular habitats. It is helpful to bear in mind the **origins of your new plants** when deciding where to put them. This will be a statement of the obvious to the experienced gardener, but the amateur often gives little thought to the matter, assuming that all but a few specialist plants commonly grown in Britain will be happy wherever they choose to plop them in the garden. If you know the origins of your plants you can make a wise choice. For instance, plants originating from forests will enjoy shade, while those traditionally found on the borders of woods and forests enjoy some sun too, and grassland varieties are happy to spend the entire day in the sun.

The British botanist E. H. **'Chinese' Wilson**, the most prolific plant hunter of all time, discovered more than 3,000 species, over 1,000 of which he brought back to Britain and the United States for cultivation. Most of those 1,000 are still grown in the West today. He survived life in the bandit-ridden wildernesses of the Far East, only to die alongside his wife when their car skidded on a wet road in Massachusetts in 1930.

Prolonging cut flowers

You can make cut **roses** last longer – as well as other flowers with hard stems – by splitting the stems and peeling back the skin or bark at the bottom. Cutting off a small amount from the stems each day also helps. Another recommended method is to stand the stems in a few inches of very hot, almost boiling water, for about two minutes before transferring them to a vase of very cold water in which you have sprinkled a dash of salt. **Tulips** also like their stems to be stood in boiling water for about a minute. **Daffodils**, meanwhile, are pretty antisocial when it comes to sharing a vase with anyone else. You should never put them in the same water as other flowers because their stems secrete a poisonous substance. To make them last longer, put the ends in an inch or two of very cold water and then leave them for three or four hours in a fridge or anywhere with a low temperature. With **narcissi**, squeeze the juice from the bottom of the stem and then lay them in cold water for up to an hour. Run cold water over them (holding them upside down) before placing in a vase. **Lilies** do well if you hold them upside down and run them under a cold tap for a minute or two before transferring them to a deep vase filled with cold water. **Poppies and peonies** should be picked just before the flowers open.

God gave us memories that we may have roses in December.

J. M. Barrie (1860–1937)

The nineteenth-century banker **Alfred de Rothschild**, a great gardening enthusiast, had over 50 glasshouses built in his garden at Halton House in Buckinghamshire. It was his gardener Ernest Field who described the status of the rich by the amount of bedding plants they set: '10,000 plants for a squire, 20,000 for a baronet, 30,000 for an earl and 40,000 for a duke.'

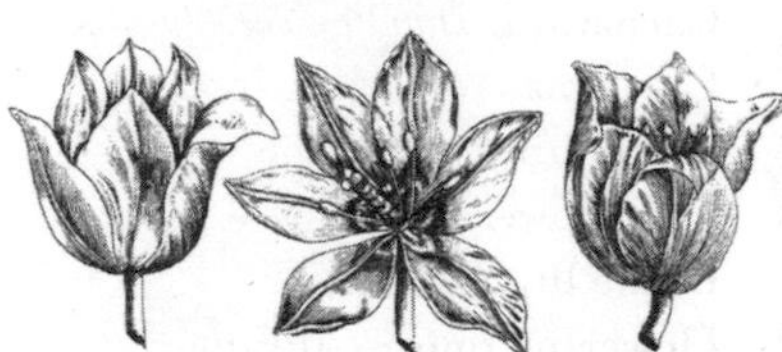

The city of Mount Vernon in Washington, USA, grows more **tulips** a year than the entire Netherlands.

The pungent smell of **garlic, parsley and onions** is highly effective for deterring greenfly and mildew on roses.

Some gardeners, garden centres and garden books often refer to plants by their botanical names, which can be confusing for the less educated gardener (i.e. me), who goes in search of flowers or shrubs that he or she knows only by their common names. Below is a list of some common flowers with both names.

African lily: agapanthus
Alpine thistle: eryngium
Snowflake/baby's breath: gypsophila
Bellflower: campanula
Bird of paradise: strelitzia
Broom: genista
Carnation: *Dianthus caryophyllus*
Christmas rose: *Helleborus niger*
Columbine: aquilegia
Cornflower: centaurea
Flame lily: gloriosa
Flowering onion: allium
French marigold: tagetes
Globe thistle: echinops
Goldenrod: solidago
Lady's mantle: alchemilla
Larkspur delphinium: consolida
Lily of the valley: convallaria
Michaelmas daisy: aster
Monkshood: aconitum
Peony: paeonia
Pot marigold: calendula
Red-hot poker: kniphofia
Snapdragon: antirrhinum
Sneezeweed: helenium
Speedwell: veronica
St John's wort: hypericum
Star of Bethlehem: ornithogalum
Sweet William: *Dianthus barbatus*
Sword lily: gladiolus
Tansy: tanacetum
Windflower: anemone
Wormwood: artemisia
Yarrow: achillea

Don't send me flowers when I'm dead. If you like me, send them while I'm alive.

BRIAN CLOUGH (1935–2004)
football manager

IN PARTICULARLY DRY, hot weather, especially when a hosepipe or sprinkler ban is in place, many plants suffer severely. One way to help **preserve the moisture** around them, in addition to laying a traditional mulch, is to surround the base of the plant with good-sized stones, gathered from elsewhere in your garden or bought from builders' merchants.

Gardens should be like lovely, well-shaped girls: all curves, secret corners, unexpected deviations, seductive surprises and then still more curves.

H. E. BATES (1905–74)
A Love of Flowers

HOPS ARE QUICK AND EASY TO grow and make for a handsome feature in the garden. You can run them along fences, over sheds or up poles. By the end of the summer the plants will be about 10 to 12 feet high, or long, depending on how you train them. (See p. 63 for uses of hops.)

The pleasure gardening in England is the article in which it surpasses all the earth.

THOMAS JEFFERSON, 1786

10 common garden flowers with strong scent

Buddleia
Escallonia
Freesia
Heliotrope
Honeysuckle
Hyacinth
Lilac
Nicotiana
Philadelphus
Rose

The word scabious comes from the Latin *scabere*, meaning 'to scratch', because the plant was thought to cure the skin disease we know as scabies.

Show me your garden and I shall tell you what you are.

ALFRED AUSTIN, 1905

10 flowers for shade

Bugle
Camellia japonica
Columbine
Cyclamen
Hellebore (Christmas rose)
Hosta 'Wide Brim'
Monkshood
Rhododendron 'Polar Bear'
Stinking iris
Variegated ground elder

FRESH FLOWERS THAT HAVE drooped, as a result of excessive heat or during transport, can be reinvigorated by placing their stems in boiling water and leaving them in a dark place until the water has cooled down. Then cut off the ends of their stems and put them in fresh water with a tablet or two of soluble **aspirin**. No kidding. A teaspoon or two of sugar or a few shots of non-diet lemonade also helps (those sachets that sometimes come with flowers from florists are basically sugar).

THE LEAVES OF THE POT marigold used to be added to **broths and stews** to give piquancy and there are still a few wise old cooks scattered across Europe who use them. The flowers were also used to deepen the yellow of butter and certain cheeses. An ointment made from the flowers to treat stings, rashes and dry skin is sold in chemists today under the name Calendula. The leaves of the pot marigold (not to be confused with tagetes, the French marigold) are said to be more effective than dock leaves for relieving nettle stings. You can make an infusion to keep at home by boiling the flowers for about 20 minutes, straining the liquid and applying it to the skin after it has cooled.

The leaves of the **Amazon water lily** (*Victoria amazonica*) reach eight feet across and can support the weight of a grown man (albeit one with very good balance). It was named after Queen Victoria following its discovery in Brazil in the nineteenth century. The ribbed structure of these mighty leaves was said to have provided the inspiration for Joseph Paxton's giant glasshouse at Chatsworth and the Crystal Palace of the Great Exhibition of 1851. The leaves grow at a rate of about one and a half square feet per day until they have reached their full size. Any other plant competing for sunlight is simply shoved out of the way or swamped by the gigantic leaves, the undersides of which are riddled with spines to prevent any passing fish from munching them.

Of the pot marigold (*Calendula officinalis*):

> *Few plants are more colourful or less fussy as to soil and situation.*
>
> William Robinson (1838–1935)

The French marigold (*Tagetes*) and bear's ear (*Primula auricula*) were brought to England by **Huguenot refugees** fleeing France after the massacre of St Bartholomew in 1572, when tens of thousands of French Protestants were killed by Catholic mobs under the sway of Catharine de' Medici, Queen of France.

CAPABILITY BROWN'S FIRST NAME was Lancelot. He was known as Capability because he was always telling his clients that their landscapes had great 'capability' for improvement.

All gardeners know better than other gardeners.

CHINESE PROVERB

THE **TULIP** HAS ITS ORIGINS IN Turkey and takes its name from the Turkish word for turban, *duliband*, in reference to its shape.

IT IS A GOOD IDEA TO **SUPPORT tall herbaceous plants** before they need it because the young plants are wasting energy when waving around in the wind, trying to thicken the cells in their stems as a means of self-defence rather than putting all their efforts into growing leaves and flowers.

THERE IS A SIMPLE, OLD TRICK to **preserve flowers** to which you may have developed a sentimental attachment (e.g. a wedding or funeral bouquet, or a bunch thrown to you on stage after a stirring performance at the Royal Opera House). Put some dry sand in a shallow box, lay the flowers on it and completely cover them with more sand. Leave in a warm cupboard or on a sunny windowsill above a radiator for a couple of weeks, after which they will be completely dry and preserved in their original shape.

Cloves are unopened flower buds.

I still cannot imagine why people do not grow these varieties [Helleborus niger and *Helleborus orientalis] more freely. They will fill up many an odd corner; their demands are few and they will give flowers at a time of year when flowers are scarce.*

VITA SACKVILLE-WEST
In Your Garden, 1951

A good rule of thumb in understanding **how much water** your plant likes is to look at the size and shape of its leaves. Those with small or needle-like leaves, such as lavender, gorse, santolina, alpines, broom and rosemary, need less water and are happy in sunny, dry positions because there is a smaller surface area from which water can evaporate. Others suitable for dry spots in the garden are those plants with fleshy leaves, built for retaining moisture, such as sedum, and plants with hairy or waxy leaves.

I perhaps owe having become a painter to flowers.

Claude Monet (1840–1926)

Agave americana is known as the century plant because it takes roughly 100 years to flower, depending on the conditions in which it is growing. It dies as soon as it has flowered. Agave is the main ingredient in the Mexican spirit tequila.

The best time to sow **wild flowers** is in mid-autumn because a number of plant varieties need the cold of winter in order to germinate effectively. It is not a disaster to sow in mid-spring, however, although some plants may not flower until the following year. It is best to choose varieties that will flower at roughly the same time, or plant two separate patches so that you get flowers in both spring and summer.

To create a little flower is the labour of ages.

William Blake (1757–1827)

Paxton's glasshouse at Chatsworth, known as the **Great Stove**, took four years to build and boasted thousands of exotic plants from around the world. It was so large that when Queen Victoria visited Chatsworth, the ancestral home of the Dukes of Devonshire, she had to be driven through the structure in a carriage. The Great Stove was enormously expensive and consumed thousands of tons of coal, the smoke of which formed dirty great clouds over the landscape. When it was decided to remove it on financial and environmental grounds, it took four attempts to blow it up.

The Ottomans are credited with creating the **language of flowers**, which attributes particular emotions and sentiments to different flowers. The writer Lady Mary Wortley Montagu is thought to have discovered the 'language' while living in Turkey in the early 1700s but it was not until her letters were published in 1759 that the phenomenon seized the imagination of first the English public, then the French and Americans. Attributes of flowers vary from source to source and country to country but the following seem to be commonly recognized:

Red rose: love
White lily: purity
Narcissi: boastfulness
Marigold: grief
Pink: boldness
Poppy: consolation
Red tulip: I love you
Nasturtium: patriotism
Heather: solitude
Buttercup: ingratitude
Anemone: forsaken

Top 10 bestselling cut flowers in the UK

1 Carnations
2 Roses
3 Lilies
4 Chrysanthemums
5 Daffodils
6 Mixed bunch
7 Tulips
8 Freesia
9 Iris
10 Sunflowers

(According to 2005 survey by the Flowers & Plants Association)

Yet the rose has one
powerful virtue to boast
Above all the flowers of the
field
When its leaves are all
dead and fine colours
are lost
Still how sweet a perfume
it will yield.

Isaac Watts (1674–1748)

PLANT HONEYSUCKLE AND clematis with their roots in the shade and their leaves and flowers in the sun.

CATNIP AND CATMINTS ARE known as such on account of the peculiar, stimulating effects they have on cats. Cats that respond to the plant love to roll in it and chew it, and sometimes they jump about making purring, growling and meowing noises. The plant valerian, used medicinally as a sedative, has the same effect on cats. (Strangely, catnip and valerian have a stimulating effect if smelled but a sedative one if eaten.) Only about two-thirds of cats are susceptible to the plant, which is related to **cannabis**, and scientists believe that the reaction is inherited. The catnip contains nepetalactone, which is thought to mimic feline sex pheromones, but some dispute whether it is this that causes the reaction as both sexes respond to it. Big cats, we are told, are extremely sensitive to catnip too, which is handy advice for gardeners if they happen to own a tiger or a puma and are umm-ing and ahh-ing about whether or not to plant it. You don't want *them* going mad in your garden, trampling all over your herbaceous borders.

THE SMALLEST FLOWER IN THE world is a type of duckweed, known as *Wolffia*, that is found on the surface of ponds. It is so small that an entire bouquet of them could fit on top of a matchstick.

FLOWER AND HERB RISOTTO

Fry some onion and garlic in olive oil, add the required quantity of risotto rice (about 150g for two people), and after a few minutes take off the heat and pour in some vegetable stock and a glass of white wine ❁ Let the mixture simmer, stirring frequently, until the right consistency is reached, and then stir in a mixture of chopped fresh herbs and flowers ❁ Let it stand for a minute or two before serving with a garnish of edible flowers (see pp. 114–15 – courgette flowers and nasturtiums work particularly well).

Dilettante gardeners love the spring and summer; real gardeners also love the winter.

ANNE SCOTT-JAMES
Down to Earth, 1971

MOST PLANTS 'SWEAT' WATER through their leaves in a process called transpiration, but **cactus** leaves have evolved into tiny spines or hairs in order to retain as much water as possible in the desert and other arid locations. The prickles serve a further purpose in stopping birds and animals from eating them. The stems of cacti are green because they have taken over from the leaves the task of making food for the plant.

THE GIANT **SAGUARO** CACTUS (the fork-shaped one you see in cowboy films) can grow up to 20m high and weigh up to six tons. Most of its weight is the water that it has greedily stored up after an infrequent shower. The saguaro, which flowers at night, can live for up to 200 years, and its fruit was a vital source of food for American Indians. Birds such as gila woodpeckers and gilded flickers make their home inside the saguaro because the temperature can be almost 30° C cooler than outside.

California produces over 60 per cent of all the fresh cut flowers grown in the United States.

Flower emblems of 20 American states

Alabama: camellia
Alaska: forget-me-not
Arizona: saguaro cactus
Colorado: Rocky Mountain columbine
Indiana: peony
Iowa: wild prairie rose
Kansas: sunflower
Kentucky: goldenrod
Louisiana: magnolia
Maryland: black-eyed Susan
Massachusetts: mayflower
Missouri: hawthorn
New Jersey: violet
New Mexico: yucca
North Carolina: American dogwood
Oklahoma: mistletoe
Texas: bluebonnet
Utah: sego lily
Vermont: red clover
West Virginia: rhododendron

A morning-glory at my window satisfies me more than the metaphysics of books.

WALT WHITMAN
(1819–91)

POOR MAN'S capers are a versatile, tangy addition to a number of dishes, including pastas, fish and salads. Capers in the shops are generally fairly expensive and this simple-to-make substitute tastes almost as good as the real thing. Poor man's capers are simply the knobbly green seeds of nasturtium, pickled in a mildly spicy vinegar. Take half a litre or so of wine vinegar and boil it with half a dozen peppercorns, the same again of cloves and a sprinkling of dried ginger. Once the mixture has cooled add the washed nasturtium seeds, seal them in jars and leave for a month or two.

Hanging baskets are happiest in a place where they get to enjoy as much morning sun as possible, but it is better if they avoid the intensity of the midday summer sun, which quickly dehydrates them.

THE **NASTURTIUM** was discovered by the Spanish in the jungles of Central and South America during the sixteenth century. It is now a commonplace sight in British gardens because it is easy to grow, striking in appearance and up there with the marigold as a companion plant for a great many vegetables, deterring a range of unwelcome insects by the pungency of its scent. Nasturtiums attract the black fly by the thousand, so it's a good idea to exploit this shortcoming by planting them near your broad beans. They are also particularly effective in protecting tomatoes, radishes, cabbages, cucumbers, squash and fruit trees.

THE WORD **NASTURTIUM** COMES from the Latin *Nasus tortus* for 'twisted nose', a reference to the peppery pungency of its taste and flavour.

Flowers are words which even a babe may understand.

BISHOP ARTHUR CLEVELAND COXE
(1818–96)

The flowers of the **evening primrose** may appear yellow to us, but insects see the ultraviolet on the petals which attracts them to the nectar.

The great British plant hunter **Robert Fortune** made four journeys to China and one to Japan in the mid-nineteenth century on behalf of the Royal Horticultural Society, the East India Company and finally the US government. He was almost single-handedly responsible for establishing the **tea industry** in India and Ceylon (now Sri Lanka) after bringing back almost 20,000 tea plants for growing in Darjeeling, thereby ending China's virtual monopoly of the market. In his quest to discover new plants, Fortune survived a great many dangers, including crazed mobs during the Boxer Rebellion, violent typhoons at sea, pirates and mountain bandits. Part of his success was based on his use of the recently invented Wardian case, a glass container which helped plants survive long journeys. Fortune also introduced several hundred new species to gardens back in England, including the plant named in his honour, *Rhododendron fortunei*.

*You can't see as well as those f***ing flowers and they're f***ing plastic.*

John McEnroe (b. 1959)
tennis champion, to a line judge

Pliny the Elder, the Roman naturalist, is credited with giving the **gladiolus** its name on account of the leaves' resemblance to the sword (*gladius*) used by Roman soldiers.

Bread feeds the body, indeed, but flowers feed also the soul.

The Koran

Echinops sphaerocephalus is the formal name for the globe thistle. Derived from the Greek, it means '**hedgehog with a round head**'.

Yes, in the poor man's garden grow
Far more than herbs and flowers
Kind thoughts, contentment, peace of mind,
And joy for weary hours.

Mary Howitt (1799–1888)
'The Poor Man's Garden'

THE AMERICAN INVENTOR **Thomas Edison** succeeded in producing rubber from goldenrod and his friend Henry Ford once gave him a Model T car with tyres made from this meadow flower.

Flowers have an expression of countenance as much as men and animals. Some seem to smile; some have a sad expression; some are pensive and diffident; others again are plain, honest and upright, like the broad-faced sunflower and the hollyhock.

HENRY WARD BEECHER
(1813–87)

20 cottage garden flowers
Achillea, buddleia, campanulas, carnations, climbing roses, cornflowers, delphiniums, foxgloves, hollyhocks, honeysuckle, lily of the valley, marguerites, marigolds, monkshood, pansies, peonies, pinks, sunflowers, sweet Williams, violas

The love of flowers is a sentiment common alike to the great and to the little; to the old and to the young; to the learned and the ignorant; the illustrious and the obscure. While the simplest child may take a delight in them, they may also prove a recreation to the most profound philosophers.

ELIZABETH KENT
Flora Domestica, 1823

10 flowers commonly found in an Elizabethan garden

Columbines
Cowslips
Hollyhocks
Lilies
Marigolds
Peonies
Poppies
Roses
Snapdragons
Sweet Williams

TO **PRESS FLOWERS** IT IS BEST TO use smaller varieties and those without thick stems or bulbous heads. Lay the flowers between blotting paper and cover them with heavy books.

Such geraniums! It does not become us poor mortals to be vain – but really, my geraniums!

MARY MITFORD
Our Village, 1820

WHEN THE FRENCH BOTANIST **Sébastian Vaillant** (1669–1722) brought up the sexual nature of plants at a lecture in Paris's Jardin du Roi, his comparisons of some plant features to the testicles and penis shocked polite society.

FOR LARGER PLANTS THAT NEED a lot of water, **submerge plastic flowerpots** next to them with the rims level with the surface of the ground and fill with water. This will drain through the pots' holes and act as a mini-watering system, feeding directly to the plants' roots. It is particularly effective in hot, dry conditions when the water runs off the surface.

WHEN YOU CUT DOWN YOUR perennials in early autumn, let them **lie in piles** in your flowerbeds and borders to provide beneficial insects (and other wildlife like toads) with a comfortable winter home. Come spring, they will be ready to go straight back to work in ridding your garden of their annoying, undesirable cousins in the insect family.

All parts of many common species of lily are highly poisonous to cats, causing kidney damage that can kill them within 24 to 72 hours, unless treated promptly. If you have a cat, check which lilies you can safely grow in the garden or have in the house.

One marked feature of the [Chinese] *people, both high and low, is a love for flowers.*

ROBERT FORTUNE (1813–80)
plant hunter

SPRING IS THE TIME TO **TAKE cuttings** from your perennial flowers, such as dahlias and delphiniums. Find shoots about two or three inches long and cut with a sharp knife, taking a little of the root as well. Put them in a mixture of compost with 20 to 30 per cent vermiculite and leave them in the greenhouse or cold frame. Once they have taken root, after about two weeks, transfer them to a larger pot of compost.

AUTUMN

Autumn vegetables
Brussels sprouts, cauliflower, onions, pumpkins, squash, sweetcorn, Swiss chard, tomatoes

What wondrous life is this I lead!
Ripe apples drop about my head;
The luscious clusters of the vine
Upon my mouth do crush their wine;
The nectarine and curious peach
Into my hands themselves do reach;
Stumbling on melons, as I pass,
Ensnared with flowers, I fall on grass.

Andrew Marvell (1621–78)

Autumn is a second spring when every leaf is a flower.

Albert Camus (1913–60)

October's the month
When the smallest breeze
Gives us a shower
Of autumn leaves.
Bonfires and pumpkins,
Leaves sailing down
October is red
And golden and brown.

Anon

Five brilliant autumn trees
Acer/maple (yellow/red/orange)
Amelanchier (red)
Gingko biloba (yellow)
Sargent cherry (rich orange)
Sweetgum (yellow, scarlet, orange)

Delicious autumn! My very soul is wedded to it, and if I were a bird I would fly about the earth seeking the successive autumns.

George Eliot (1819–80)

The goldenrod is yellow
The corn is turning brown
The trees in apple orchards
With fruit are bending down.

Helen Hunt Jackson (1831–85)

So dull and dark are the November days.
The lazy mist high up the evening curled,
And now the morn quite hides in smoke and haze;
The place we occupy seems all the world.

John Clare (1793–1864)

Autumn Birthdays

16 September 1812: Robert Fortune, Duns, Berwickshire
18 October 1616: Nicholas Culpeper, London
31 October 1620: John Evelyn, Wotton, Surrey
29 November 1843: Gertrude Jekyll, London

Season of mists and mellow fruitfulness,
Close bosom-friend of the maturing sun;
Conspiring with him how to load and bless
With fruit the vines that round the thatch-eves run;
To bend with apples the moss'd cottage-trees,
And fill all fruit with ripeness to the core;
To swell the gourd, and plump the hazel shells
With a sweet kernel; to set budding more,
And still more, later flowers for the bees,
Until they think warm days will never cease,
For Summer has o'er-brimm'd their clammy cells.

JOHN KEATS, 'To Autumn', 1820

Everyone must take time to sit and watch the leaves turn.

ELIZABETH LAWRENCE (1904–85)

No spring nor summer beauty hath such grace
As I have seen in one autumnal face.

JOHN DONNE (1572–1631)

No warmth, no cheerfulness, no helpful ease,
No comfortable feel in any member — No shade, no shine, no butterflies, no bees,
No fruits, no flowers, no leaves, no birds — November!

THOMAS HOOD, 'No!', 1844

The foliage has been losing its freshness through the month of August, and here and there a yellow leaf shows itself like the first gray hair amidst the locks of a beauty who has seen one season too many.

OLIVER WENDELL HOLMES
(1809–94)

Fruit

Of the strawberry: *Doubtless God could have made a better berry, but doubtless God never did*

It is remarkable how closely the history of the apple tree is connected with that of man.

HENRY DAVID THOREAU (1817–62)
American author and naturalist

THE SEEDS OF APPLES, ALMONDS, apricots, cherries and peaches all contain **cyanide**. In apples, the quantity is so small that you would have to eat bags of crushed pips to be fatally intoxicated. The amount of cyanide in peach and apricot pits, however, is potentially very harmful, but fortunately they are so large and hard that few people swallow or chew them.

The apple can grow at the highest latitude of all the fruits. Apple trees need almost 1,000 hours of cold autumn and winter air to produce their flowers, which explains why they cannot be grown in hotter climates.

APPLE CHUTNEY

Take about two kilos of chopped and peeled apples, half a kilo of chopped onions, a few crushed garlic cloves, half a kilo of sultanas and the same again of soft brown sugar, half a litre of malt or wine vinegar, a couple of teaspoons of ground ginger and some seasoning ✿ Throw all the above into a large pan, bring to the boil and then simmer for four or five hours, stirring occasionally ✿ Transfer into warm jars when still hot.

IT TAKES THE ENERGY FROM **50 leaves** to grow one apple.

THE **NATIONAL FRUIT Collection** at Brogdale Farm, Kent, has gathered a vast amount of different varieties of fruit grown in temperate climates with the aim of conserving genetic diversity. At the time of writing the 'living' collection, which has become an important international resource, comprises the following numbers of varieties:

Apple 2,040
Apricot 11
Blackcurrant 109
Cherry 322
Cider apple 102
Cobnut 48
Gooseberry 144
Medlar 5
Ornamental malus 80
Ornamental prunus 48
Ornamental pyrus 5
Pear 502
Perry 20
Pinkcurrant 4
Plum 337
Quince 19
Redcurrant 73
Vines 51
Whitecurrant 20

THE **APPLE** WAS THE EARLIEST OF all fruits to be cultivated by man and, like so many other plants, it was introduced to Britain by the Romans. Today, China is the world's leading apple producer.

Eat an apple going to bed
Knock the doctor on the head.

ANON

THOUSANDS OF VARIETIES of apples have been grown in Britain over the years, making it all the more **unbelievable and regrettable** that only a handful are readily available in shops today, the majority of which have been flown in from around the world, at great cost to the environment and local growers, not to mention our taste buds (supermarket demand has meant that foreign growers are more concerned with quantity and neat appearance than taste and nutritional properties).

Over 60 per cent of **Britain's apple orchards** have disappeared in the last 50 years, many as a result of ridiculous financial incentives encouraging growers to grub up their trees. In 1987 there were 1,500 registered apple growers; today there are fewer than 500. Roughly 68 per cent of the apples sold here are foreign imports. Do your bit for the British apple and apple grower by voting with your shopping basket.

THE STRAWBERRY IS NOT strictly a fruit, but a member of the rose family. What you eat is the enlarged receptacle at the end of the flower. Each strawberry contains an average of **200 tiny pips**, which are the true fruits.

STRAWBERRIES USED TO BE used to whiten teeth and remove plaque. The practice of cutting them in half and rubbing them on teeth was especially popular with the Swedes. Today dentists say the acid could do as much harm as good, but it still works well if you brush with fluoride straight afterwards. Rubbing your teeth with a sage leaf is another good way of whitening them.

There are two or three objects which you should endeavor to enrich our country with. One is the Alpine strawberry.

THOMAS JEFFERSON
letter to James Monroe, 1795

TODAY MOST PEOPLE MULCH their strawberries with straw, but you can improve their growth and taste by mixing in an equal amount of **pine needles**. It is better still if you can plant them in soil removed from around pine trees.

NINETY-FIVE PER CENT OF Britain's blackcurrant harvest goes towards the production of cordials. The future of the drinks industry in its traditional heartland of the West Midlands is threatened by our steadily warming climate. **Currants need a cold period** to grow successfully, and if Britain becomes much warmer, growers may be forced to move northwards or seek alternative produce. According to the Department for Environment, Food and Rural Affairs (DEFRA), the production of a whole host of other fruits could also become commercially unviable by 2050, with the apple, pear, strawberry, raspberry and rhubarb all under threat. They, too, need a long cold period to flower and fruit.

STRAWBERRIES WERE CONSIDERED an aphrodisiac in the Middle

Ages and a soup made of strawberries, borage and cream was traditionally served to newly-weds for breakfast.

Youth is like spring, an over-praised season more remarkable for biting winds than genial breezes. Autumn is

the mellower season, and what we lose in flowers we more than gain in fruits.

SAMUEL BUTLER (1835–1902)

Raspberries can be grown anywhere from the Arctic to the Equator.

PEACH MELBA AND **Melba sauce** were created by the famous chef Escoffier and named in tribute to the Australian soprano Dame Nellie Melba. To make this simple sauce, add half the weight of your raspberries in icing sugar and pour it all into a blender for a few seconds, then sieve the mixture to get rid of the pips.

BLACKCURRANTS WILL grow very happily and vigorously if planted in an old nettle patch. If your blackcurrants are already established, you can improve their quality by planting some nettles among them.

BLACKCURRANTS CONTAIN more than four times as much vitamin C as oranges.

TO PROTECT YOUR SOFT FRUIT from **irksome, greedy birds**, wind and tie some thin black thread between the branches. It is all but invisible to the human eye but birds find it hard to pick out the thread and will not risk getting trapped in it. (This trick also works on seed beds if you tie thread to a few small sticks zigzagged along the row.) Tying old CDs to the branches will also deter birds.

SLOES ARE THE PURPLY black fruit of the blackthorn tree. To make deliciously flavoured gin from them (you can also use vodka if you're feeling a bit racy and modern), pick the sloes in October or November when they are most ripe after the first frosts. Cut the sloes and drop them into a half-full bottle, adding a teacup's worth of sugar. The bottle will need to be turned once a day for the first week, then once a week for the first month or two. You can start drinking it after two months, but it is best if left for a year. You can make similar concoctions if you have an excess of **damsons or juniper berries**. For years, sloes were also used to produce a very strong dark-red dye for clothing.

ANOTHER GOOD WAY OF protecting fruit trees from insect attack is to plant lavender close by, or train some climbing nasturtiums up the trunks. This, however, will be no protection against **wasps**, which will happily eat your fruit all day long if you show them the green light. One way of deterring wasps is to provide them with a sweeter alternative to that which you are trying to protect. Old jam jars filled with beer or watered-down jam or honey, leaving enough room for the creatures to drown in, has passed the test of time.

The wasps which are without number this dry hot summer attack the grapes in grievous manner. Hung up 16 bottles with treacle and beer which make a great havock among them . . . Mr Snooke's grapes were eat naked to the stones a fortnight ago, when they were quite green.

GILBERT WHITE
The Natural History and Antiquities of Selborne, 1762

NEW ZEALANDERS JOKE that there are two types of **blackberry**: one in the North Island and one in the South. (In Europe there are 300 wild species.) In the Middle Ages their sinister reputation meant that they were often allowed to grow freely in churchyards to keep the ghosts of the deceased in their coffins.

BLACKBERRY AND APPLE CRUMBLE

There are three good reasons why this traditional British recipe has been so popular down the centuries: 1) it's bloody delicious 2) it's incredibly easy to make 3) it is an excellent way of using up all those dozens of cooking apples and thousands of blackberries that would otherwise just rot ✿ Stew about half a kilo of chopped and peeled apples with about 50g of sugar and two tablespoons of water until they are half cooked, then add about 250g of blackberries, and pour the whole lot into a pie dish ✿ For the crumble, use your fingers to combine a big knob of butter with about 150g of plain flour and about 75g of caster sugar, and spread it all over the fruit stew ✿ Put it in a medium-hot oven for about 40 minutes.

And the eyes of them both were opened, and they knew that they were naked; and they sewed fig leaves together and made themselves aprons.

Genesis 3:7

[The snail] *is very mischievous, and especially among fruit-trees, where it annoys the fruit, as well as the leaf, but particularly the fruit. It is a great enemy of the apricot and the plum.*

William Cobbett (1763–1835)
The English Gardener

Figs, together with olives, helped make Athens rich. They were also a staple food of Sparta and athletes ate them in great quantities for strength. The fruit was so popular that their export had to be regulated by the authorities, and it is from this ancient law that we have the English word **sycophant**. *Sukophantes* (from *sukon*, meaning 'fig' and *phainein* meaning 'to show') was the Greek word for an informer, i.e. someone who reported an illegal fig exporter.

It has been argued by some historians in recent times that the **decline of the Roman Empire** had less to do with moral degeneration and more to do with the way the Romans of the ruling classes made their **wine**. While the poorer classes

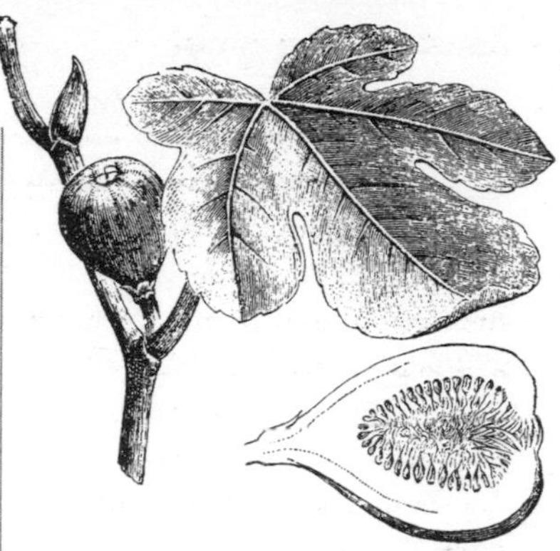

used earthenware pots and animal skins, the rich used lead pots and kettles (as well as for cooking and in their plumbing), leading to chronic poisoning, which by turn brought on anaemia and weight loss. There are historians who disagree with the theory but it is certainly true that the birth rate of the ruling classes went into decline in the later years of the Empire and more of them began to die at a young age.

Healthy, **mature apple** trees will bear between 100 and 200 kilos of fruit every year.

The leaves of **rhubarb**, whether cooked or raw, contain enough oxalic acid to kill a man, but its dried and powdered roots have been used for centuries to treat a number of intestinal complaints, including **amoebic dysentery**.

With their close standing we find vast Numbers of Orchards that have scarcely a healthy Tree in them, the greatest part of them being either cankered, or covered over with Moss; and how can we suppose to eat kindly Fruit from distempered Trees?

PHILIP MILLER
Scottish-born gardener and botanist, head gardener at Chelsea Physic Garden from 1722 to 1770

APPLES AND PEARS ARE **EASY TO store**, demanding only a box or shelf to sit on in a cool, dark, slightly damp environment (an old church would be ideal but, assuming you don't enjoy such a feature in your garden, the garage will be fine). It is advisable to wrap each apple in newspaper to prevent mould. Make sure you store only the best fruit, removing all dubious-looking specimens which might ruin the healthy ones. It is also important that air can circulate around the fruit as poor ventilation will cause rotting.

MOST JAMS ARE FAIRLY **EASY TO make**, and the general principle that you boil fruit with the same weight in sugar holds good. Plum jam will rarely fail because the high amount of pectin in the fruit makes it perfect for setting. If your plums are especially sweet, then you might want to reduce the amount of sugar. De-stone the plums and for best results cut them in half, sprinkle with sugar and leave overnight. For every kilo of fruit, you will need about 350ml of water, Simmer until you have a thick, tender mess, then add the sugar and boil rapidly for about 10 minutes or until the mixture starts to set. It is often difficult to get strawberry and raspberry jam to set because they are low in pectin, but you can overcome this by **adding lemon juice**, which helps draw out the pectin. Alternatively, you can boil up separately some slices of lemon, skin, pith and all, and then pour the liquid into the jam. (Pectin is also sold in ready-prepared powder form.) Another way round the pectin/gelling problem is to combine redcurrants, which are high in the stuff, with fruits that are low such as pears and rhubarb.

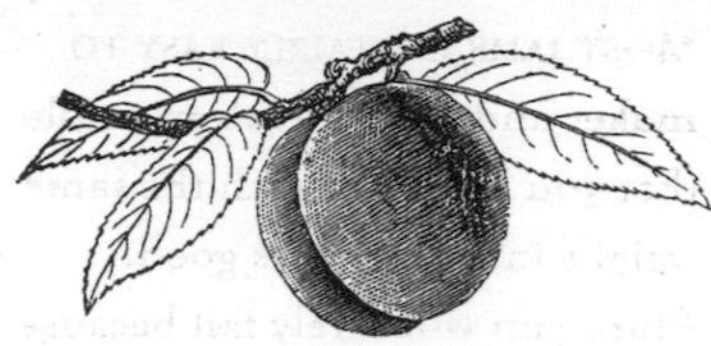

The way to keep birds from fruit, and, indeed, from everything else, is to shoot them, or frighten them away.

WILLIAM COBBETT (1763–1835)
The English Gardener

FOR CENTURIES **RASPBERRY LEAF tea** has been used effectively to treat diarrhoea and as a gargle to soothe sore throats. Making it is even simpler than brewing an everyday pot of tea: take an ounce of dried raspberry leaves and infuse them in hot water.

DRIED FIGS CONTAIN ABOUT 60 per cent sugar, making them an excellent energy snack.

The sun, with all those planets revolving around it and dependent on it, can still ripen a bunch of grapes as if it had nothing else in the universe to do.

GALILEO (1564–1642)

MULBERRIES SHOULD BE EATEN as soon as possible after picking because they won't last more than a day or two, even in the fridge, before they start to grow mould. (It is for this reason that they are such a rare sight in our supermarkets and greengrocers.) Mulberries are delicious eaten fresh but they are also good stewed and as they have such a high water content you can cook them in their own juice. They will soon turn into a **sweet runny mess** – good for pouring over vanilla ice cream or mixing with natural yoghurt. You can add a squeeze of lemon juice or some orange zest to reduce the sweetness.

AN EFFECTIVE ORGANIC WAY OF killing off **greenfly** on your roses: boil up some **rhubarb leaves**, which contain high quantities of poisonous oxalic acid, and then spray on the cooled liquid.

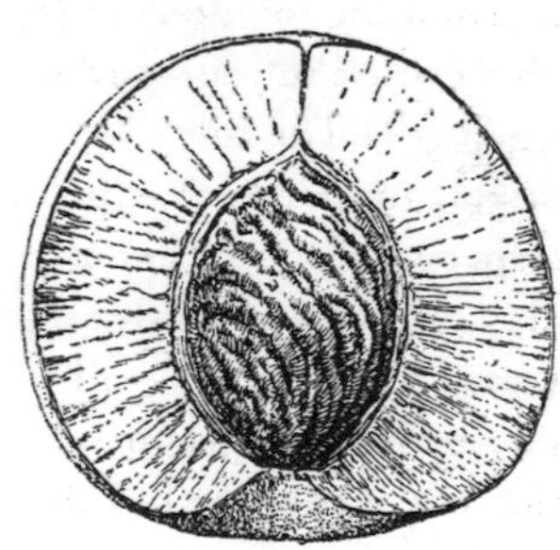

BRITAIN IS THE ONLY COUNTRY that grows **apples** expressly for cooking.

You need roughly 35 to 40 apples to make eight pints of cider.

BRITAIN REMAINS THE WORLD'S chief cultivator of the **gooseberry**. In the mid-eighteenth century Lancashire weavers, recently arrived in the towns from their cottages in the countryside, took to growing gooseberries in pots in the absence of any land or gardens to cultivate. Gooseberry competitions, to see who could grow the largest berries, became a popular pastime and a great number of **gooseberry clubs** were formed across the north in the years that followed.

BRITONS HOLD THE **WORLD records** for growing three of the heaviest fruits: Alan Smith for his 1.67kg apple, Kelvin Archer for his 61g gooseberry and a certain G. Andersen for his 231g strawberry.

FRUIT MESS

If you have a pile of soft fruits that you cannot eat fresh – or don't want to – at a single sitting, there are a number of ways of making good use of them ❁ Simmer the fruit in a large pan with some sugar and a small amount of water until the mixture is soft ❁ You can now eat it in four different ways: (1) as the simple fruit stew that you see before you, hot or chilled (2) as a purée (i.e. sieved to remove seeds and skins (3) as a fool – just stir in double cream or natural yoghurt and leave to chill in the fridge (4) in homemade ice cream.

DEVOTEES OF RHUBARB WILL tell you that the variety 'forced' in the winter months is superior to that which we find in our gardens a few months later. Large-scale or commercial forcing is carried out indoors and in total darkness, where the shoots, aided by the warmer environment, race upwards in a frantic search for light. Rhubarb grows so fast in these conditions that the plants make a popping noise when the buds burst into leaf. The British rhubarb-growing industry began in the 1880s, centred on a tiny triangle in West Yorkshire bordered by Wakefield, Leeds and Bradford. Such was the demand that by the 1950s there was a 'rhubarb express train', carrying tons of the fruit (strictly a vegetable, but let's not argue) to customers in the south of the country. Today, the glut of more 'exotic' fruits from abroad has led to a collapse in demand, and there are now just a dozen growers where once there were over 200. Radio 4's *Today* programme, however, recently reported a renewed interest in the fruit in the country's fancier restaurants.

AMERICANS STARTED USING **blueberries** in their muffins when they discovered that the fruits exploded on reaching a certain temperature and distributed their flavour throughout the sponge.

Noah was the first tiller of the soil. He planted a vineyard; and he drank of the wine, and became drunk, and lay uncovered in his tent. And Ham, the father of Canaan, saw the nakedness of his father, and told his two brothers outside. Then Shem and Japheth took a garment, laid it upon both their shoulders, and walked backward and covered the nakedness of their father; their faces were turned away, and they did not see their father's nakedness.

GENESIS 9:20–3

TORTURERS IN THE MIDDLE Ages, ever alert to more and more hideous means of inflicting pain, were delighted by the development of a very nasty little tool that came to be known as the 'Pear of Confession'. The implement, shaped like a pear, was inserted into the mouth and then slowly expanded.

Have you ever wondered how they manage to put a fully grown pear inside a bottle? You can make your own Poire William-style liqueur, simply by placing an empty bottle over a young bud and tying it to the branch. When the pear is nearly ripe, pick and fill the bottle with spirit, preferably some that has been distilled from fruit.

BLACKBERRIES AND APPLES ARE the two fruits in your garden you are most likely to have a surfeit of, come the autumn. If you have had your fill of crumble, you can turn them into a jam with a simple recipe. You need roughly double the amount of blackberries as apples, so stew a kilo of berries in one pan and half a kilo of apples in another with about 100ml of water in each. When both are soft, mix them in the larger pan, add about 1.5 kilos of sugar and boil, stirring frequently, until the mixture starts to set. Like currants and plums, apples are high in pectin which helps the jam to set easily. It may set within a few minutes but it can take up to 15. If you are unsure, take a teaspoon of the liquid and place it on a plate. If a skin forms then it has set. Transfer the jam while still hot into (warmed) jars and add a disc of waxed paper to the surface before sealing the jar tight.

IF YOU STILL HAVE BILLIONS OF blackberries in your garden that even the birds have tired of devouring, and you cannot bear the thought of them decaying uneaten, then you can turn them into a **delicious cordial** that will last for a long time if you keep it bottled in a cool dark place. It's cheap, easy and more natural than the manufactured varieties. Cover the blackberries with white wine vinegar, cover the bowl and leave them for about a week, stirring at least once or twice a day. Strain the mixture and boil it hard, adding about half a pound of sugar for every pound of blackberries, and also about a quarter pound of honey. Let it cool and then bottle.

THE PHRASE '**GONE PEAR-shaped**' has its origins in the officers' mess of the Royal Air Force, and was used as a humorous way to describe the shape of an aeroplane that had crashed nose first.

THE **CRABAPPLE FAIR** IN Egremont on the west coast of Cumbria claims to have been staged every year (interrupted only by two world wars) since a Royal Charter was granted in

1267. The fair, which is held every autumn, traditionally features a bizarre series of events, including a challenge to climb a greasy pole to claim the joint of mutton fixed to the top. A lorry (it used to be a cart) is driven through the streets of the town while men distribute apples to the crowd. There is also a **pipe-smoking contest** and something called the World Gurning Championship, in which competitors put their heads through a horse collar and pull the most ridiculous faces possible. The winner is the person who receives the most applause.

Of the strawberry:

Doubtless God could have made a better berry, but doubtless God never did.

DR WILLIAM BUTLER (C.1536–1617)

TO **PREVENT BLIGHT**, HANG dead tomato plants from the branches of an apple tree or burn them beneath it.

The wealthy people [in Virginia] *are attentive to the raising of vegetables, but very little so to fruits. The poorer people attend to neither, living principally on a milk and animal diet. This is the more inexcusable, as the climate requires indispensably a free use of vegetable food, for health as well as comfort.*

THOMAS JEFFERSON, 1782

The peach is native to China and is thought to have reached the Mediterranean region in roughly 2000 BC via the Silk Road.

THERE ARE A NUMBER OF time-honoured tips for protecting your fruit from insect pests, predators and disease. **One is burning garden refuse** under the branches, which acts as a good fumigant to flush out unwelcome insects.

FIGS ARE HIGHLY SUITED TO growing in containers because pots restrict root growth, which is essential for the production of good fruits. When growing figs in the open ground, you can contain the roots of the tree by inserting paving slabs into the hole prior to planting. Figs originate in warmer climates than ours and it is important that they are protected from frost in the winter. If you have the space, move them indoors to a greenhouse, shed or garage, but if not, cloak them in bubblewrap and fleece stuffed with straw.

Trees and Shrubs

A garden without trees scarcely deserves to be called a garden

The crabapple is the only apple indigenous to Britain.

GINGKO BILOBA, ALSO KNOWN AS the maidenhair tree, a living fossil dating back 300 million years, is a unique tree in a genus all of its own, without even distant relatives. Very tough trees, generally free from disease and impervious to most common pests, they can grow almost anywhere and are especially good in cities because they are not in the least bit bothered by pollution. The gingko survived the atomic blast in Hiroshima. Four were situated near the blast centre and when they were examined a few weeks later they were starting to bud. Two are still alive today.

HEALTHY FRUIT TREES NEED AIR to be able to circulate freely round their branches because it prevents disease and helps ripen the fruit. Trees therefore should be planted a **reasonable distance** apart from each other, generally between eight and ten yards. For the same reason they also need to be pruned once a year. And it is a good idea to thin the crop early in the summer.

CRABAPPLE JELLY

Cut up the crabapples and stew them until they are a thick mush ❁ You can add a few cloves, a pinch of cinnamon and/or some small pieces of ginger for extra flavour or piquancy ❁ Strain the stew overnight by tying it up in a muslin cloth and suspending it over a pan into which it will slowly drip ❁ Add roughly the same quantity of sugar (ready dissolved preferably) and boil the mixture hard until it sets (roughly 10 to 15 minutes) ❁ Transfer to clean, warmed jars, place a disc of greaseproof paper on top and seal.

Native British Trees

Alder
Ash
Beech
Black poplar
Cherry
Crabapple
Elder
Elm
Hazel
Holly
Lime
Oak
Rowan
Scots pine
Silver birch
Yew
Willow (cricket bat variety)

THE SCOTS PINE IS THE OLDEST and most common conifer in Britain, and it is the only pine in northern Europe to have withstood the last Ice Age. The Scots pine, the yew and the juniper are the only needle-leaved evergreens native to our shores. The needles of pine trees allow them to retain moisture, which makes them excellent trees to plant in dry or sandy locations. For centuries the English navy made extensive use of the pine tree's tar to seal their wooden ships, and such was the demand for the substance dubbed '**black gold**' that it was also imported in great quantities from Sweden. Vets and farmers also used the tar as a highly effective antiseptic for livestock, while brewers and coopers sealed the inside of their barrels with it.

THE **BULLFINCH**, ONCE ONE of our most abundant birds, has a voracious appetite for young fruit-tree shoots, and can remove around 50 buds a minute. The attacks became so bad in the 1950s that they caused a crisis in Britain's fruit-growing industry and led to growers trapping and killing the birds in their hundreds of thousands each year. Their numbers have reduced sharply in recent decades, coinciding with the introduction of the sparrowhawk in the 1970s.

CAPTAIN COOK AND HIS CREW are thought to have given the tea tree its name after using the leaves to make a hot infusion when they came ashore in New South Wales.

A
NEW ORCHARD and Garden.
OR
The beſt way for planting, grafting, and to make any ground good, for a rich Orchard: Particularly in the North parts of England: generally for the whole kingdome, as in nature, reaſon, ſcituation, and all probability, may and doth appeare.
WITH
The Country Houſewifes Garden for hearbes of common vſe, their vertues, ſeaſons, profites, ornaments, variety of knots, models for trees, and plots for the beſt ordering of Grounds and Walkes.
AS ALSO.
The Husbandry of Bees, with their ſeuerall vſes and annoyances, all grounded on the Principles of Art, and precepts of Experience, being the Labours of forty eight yeares of William Lawſon.

Printed at London by Bar: Alſop for Roger Iackſon, and are to be ſold at his ſhop neere Fleet-ſtreet Conduit. 1618.

What can your eye desire to see, your ears to hear, your mouth to take, or your nose to smell, that is not to be had in an orchard, with abundance of variety?

WILLIAM LAWSON
A New Orchard and Garden, 1618

ELDERFLOWER 'CHAMPAGNE' AND CORDIAL

By the end of May, the elder tree is in full blossom, and you can seize this natural window of opportunity to make a stock of delicious cordial to last throughout the summer ✿ Take roughly two dozen to 30 elderflower heads and put them in a large bowl with a couple of sliced lemons, a kilo of sugar, 30g of citric acid (from your chemist) and pour over two litres of boiling water (don't wash the flowers as that gets rid of much of the flavour) ✿ As an alternative to citric acid, use more juice from a couple of lemons and also add some peel ✿ Cover the bowl and leave in the fridge, or a cool corner of a garage or outhouse, for five days before sieving the cordial and bottling what you need for a week or so ✿ Freeze the rest in small quantities as it has no preservatives and will go off after a week or 10 days ✿ To drink, put a finger of the cordial into a glass and fill up with sparkling water ✿ (N.B. Citric acid is also used to help the injection of heroin so don't be alarmed if chemists ask why you want it!)

Like so many traditional foodstuffs of Britain, **sweet chestnuts** have slowly disappeared from our tables, which is a shame because they are delicious, wholesome and plentiful. They are particularly good in a soup, which is so easy to make you could get the **village idiot** to do it. Melt some butter in a pan, add some chopped garlic, onion and celery, and cook till soft. (You can also add some sliced potato if you want more bulk.) Make a nick in your chestnut shells (otherwise they'll explode) and bake your chestnuts for 10 minutes, then peel them while they're hot and add the nuts to the onion mix with a pint or so of stock, depending on how many you are cooking for. Let it simmer for 10–15 minutes, and then whiz in a food processor. Many recipes recommend sprinkling on some nutmeg and adding a dash of cream when you serve it up.

Dried **sweet chestnuts** need to be soaked for at least 90 minutes and then boiled for about an hour before you peel them. If you preserve them in syrup, you will have made a basic form of the famous French delicacy marron glacé.

Except during the nine months before he draws his first breath, no man manages his affairs as well as a tree does.

George Bernard Shaw
(1856–1950)

The berries [of the bay tree] *mightily expel the wind, and provoke urine, help the mother, and kill the worms.*

Nicholas Culpeper
The Complete Herbal, 1653

The leaves of the **laurel** tree were made into wreaths and crowns in Roman times for military heroes and poets, hence the title 'poet laureate', meaning 'the wreathed or crowned poet'. Traditionally, laurels were planted near houses to ward off witches and protect the house from lightning.

Oaks are among the **slowest growing** trees and usually don't produce acorns until they are about 20 years old.

Tis thought the king is dead; we will not stay.
The bay trees in the country are all wither'd.

William Shakespeare (1564–1616)
Richard II

Water travels up tree trunks at roughly 150 feet per hour.

The **mulberry tree** is usually the last to come into leaf and traditionally country gardeners took it as a sign that they could not plant out free from the risk of a late frost until the buds of the mulberry had opened. Older gardeners still call the mulberry 'the wise tree'.

When mulberry trees are green
No more frosts are seen.

Old saying

Mulberry trees have been known to live for up to 600 years. Syon House in Brentford holds the two known oldest in Britain today, both planted in 1548. Built for survival, if they are blown over in a storm they soon grow back again by raising themselves up on their branches.

With time and patience the mulberry leaf becomes a silk gown.

Chinese proverb

In order to prevent witches from entering their houses, the common people used to gather Elder leaves on the last day of April and affix them to their doors and windows.

William Cole
The Art of Simpling, 1656

A garden without trees scarcely deserves to be called a garden.

Canon Henry Ellacombe
In a Gloucestershire Garden, 1895

To guarantee berries on your Christmas **holly sprigs** and wreaths, cut them about two or three weeks before you put up your decorations, and then hang them up in a shed or garage where the mice can't get at them. If you leave the cutting to the last minute, you can be fairly certain that the birds will have stripped off all the berries. If you are wondering why your holly tree or bush never produces any berries, it may well be because you have only one. Although there is a self-fertilizing variety of holly, you need two trees or bushes of most varieties, one male and one female, in order to produce berries.

The world's tallest topiary has been created by a gardener and agricultural scientist called Moirangthem in Manipur, India. According to the *Guinness Book of World Records* Moirangthem calls his giant plant (the Duranta hedge plant) 'sweetheart'. On the day his achievement was recognized, his friends celebrated by lighting candles.

Elder is perhaps the most neglected hedgerow plant, though it is the source of elderflower wine, of cordial, and also of an excellent pudding – the flowers deep-fried in batter and sprinkled with sugar are a great delicacy. Elders keep witches at bay, and any old cottage that has not been modernised will have one growing alongside.

Derek Jarman
Derek Jarman's Garden, 1995

10 plants for hedges

Beech
Blackthorn
Bramble
Crabapple
Hawthorn
Hazel
Holly
Hornbeam
Wild privet
Wild service

THE HORSE CHESTNUT TREE, introduced to Britain in the early 1600s, is a native of Macedonia and Albania. There are two possible explanations for its name. One theory claims it derives from the fact that when you pluck one of the leaves the shape of a horseshoe remains at the end of the stalk. Another theory is that for many centuries Turkish horsemen

treated bruises to their beasts with aescin, a natural chemical obtained from the fruits. The Turks were also said to have fed the chestnuts to their horses to stop them farting. During the two world wars, horse chestnuts were processed for their starch content to manufacture acetone, needed in the production of cordite for armaments. Today, herbalists use extracts of chestnut for the prevention of haemorrhoids, while aescin is also used in the treatment of varicose veins, sprains and edema.

Under the spreading chestnut tree
The village smithy stands;
The smith, a mighty man is he,
With large and sinewy hands,
And the muscles of his brawny arms
Are strong as iron bands.

HENRY WADSWORTH LONGFELLOW (1807–82)
'The Village Blacksmith'

UNTIL ABOUT 6,000 YEARS ago Britain was almost entirely covered in **woodland**, but it has been gradually cleared for farming and for building communities. Even the Highlands of Scotland, constantly described as a 'wilderness', are in fact an artificial landscape created by man, who has systematically removed the ancient fir forests over the centuries. Woodland now covers just over 11 per cent of the British landscape and just 2 per cent of ancient woodland (i.e. that which has existed since pre-1600) survives.

Human urine is an excellent source of nitrogen for plants, and is also handy for accelerating the composting process.

There will not be a forest tree left standing in the whole Kingdom.

Sir William Chambers (1723–96)
fellow landscape architect and fierce critic of Capability Brown, whose lowly origins he also enjoyed mocking

Over 5,000 mature oak trees, covering 80 acres of forest, were needed to build one ship of the line in Nelson's navy. Once felled, the trees had to be stored for years so that the wood seasoned properly. They were then cut, bent over a fire pit, doused with water and pressed with heavy weights until the right shape had been achieved. The trunks of seven large elm trees were used to make the keel of Nelson's HMS *Victory* and roughly 2,800 fir and spruce trees went into the decks, masts and yardarms.

The trees generally lose their leaves in the following succession: walnut, mulberry, horsechestnut, sycamore, lime, ash; then after an interval, elm; then beech and oak; then apple and pear trees, sometimes not till the end of November; and lastly pollard oaks and young beeches, which retain their withered leaves till pushed off by their new ones in spring.

Leigh Hunt
The Months, 1821

Elm wood is waterproof and was used throughout Europe for making pipes during the Middle Ages.

DAMSON PORT

Pour five litres of boiling water over roughly two kilos of damsons and leave them for about 10 days to two weeks, stirring the mixture and squeezing the fruits at least once a day ❁ Strain the mixture through a muslin cloth or jelly bag hung over a large pan and then strain twice more without squeezing ❁ Add two kilos of dissolved sugar to the damson mixture and pour in half a pint of boiling water to raise the temperature ❁ Leave to ferment for two or three weeks and after skimming the top of debris, pour into bottles using a funnel, and cork.

In a single season a large **elm** tree makes about six million leaves.

Apples be ripe, nuts be brown
Petticoats up, trousers down.

English saying

SILK COMES FROM THE **SILKWORM**, whose diet consists entirely of the leaves of the mulberry tree. In an effort to establish a thriving silk industry in England, James I issued an edict in 1608 encouraging the planting of thousands of mulberry trees throughout the land. The venture proved unsuccessful, it is thought, because it was the black mulberry and not the white mulberry, eaten by the silkworm, that was planted. The white mulberry, it later turned out, struggles in our damper climate.

He who plants a tree, plants a hope.

OLD SAYING

THE SLOE-PRODUCING **blackthorn** tree usually appears in the form of a shrub, but it will turn into a tree with a recognizable trunk if it grows in the shade of taller trees.

British gardeners, instead of humouring nature, love to deviate from it as much as possible. Our trees rise in cones, globes and pyramids. We see the marks of scissors on every plant and bush. I do not know whether I am singular in my opinion but, for my own part, I would rather look upon a tree in all its luxuriancy and diffusion of boughs and branches, than when it is thus cut and trimmed into a mathematical figure, and cannot but fancy that an orchard in flower looks infinitely more delightful than all the little labyrinths of the most finished parterre.

JOSEPH ADDISON (1672–1719)
essayist, in the *Spectator*

THE SCOTTISH BOTANIST AND plant hunter David Douglas, after whom the **Douglas fir** is named, died in extraordinary and terrifying circumstances. On his return from the north-western United States, his ship docked in Hawaii to replenish supplies and Douglas seized the opportunity to explore the mountains in search of new species. Tragically, he fell into a pit-trap dug to capture ferocious wild cattle that roamed the mountainsides. One such beast was already in there when he fell in and Douglas was powerless to prevent it trampling and goring him to death. His remains were discovered by a search party a few days later.

The pine tree seems to listen, the fir tree to wait: and both without impatience – they give no thought to the little people beneath them devoured by their impatience and their curiosity.

FRIEDRICH NIETZSCHE
(1844–1900)
The Wanderer and His Shadow

BEECH LEAVES CAN BE preserved with glycerine for attractive, long-lasting winter arrangements. Cut the sprays in summer while the sap is still rising and the leaves have yet to turn brown. Split the stems with a sharp knife and stand them up in a container with equal parts of water and glycerine. Leave them for a couple of weeks. The leaves will turn a glossy brown colour and can be used without water.

THE PERSIAN WORD FOR enclosure was *pairidaeza*, which became *paradeisos* in Ancient Greek and eventually *park* in English.

The tree which moves some to tears of joy is in the eyes of others only a green thing that stands in the way. Some see Nature all ridicule and deformity, and some scarce see Nature at all. But to the eyes of the man of imagination, Nature is Imagination itself.

WILLIAM BLAKE
Letters
1799

12 trees to grow in containers

Apple
Box
Cherry
Citrus
Fig
Holly
Nectarine
Olive
Peach
Pear
Plum
Yew

(Citrus, fig and olive trees will have to be moved indoors or protected from frost in the winter.)

Birch trees don't hang around long compared to many of their leafy cousins, and will have done well to reach 100 years old by the time they give up and topple over. A birch tree has shallow roots, and when it is blown over in a gale the wood rots but the bark doesn't, leaving a long hollow tube. The bark's paper-like quality made it perfect as an early form of parchment, as well as excellent kindling in snowy or damp conditions. Today birch trees are still grown in great quantities for plywood.

A society grows great when old men plant trees whose shade they know they shall never sit in.

Greek proverb

Five trees to grow that make a particularly nice sound in the wind

Aspen
Bull Bay magnolia
Eucalyptus
Fig
Paper birch

10 trees for small gardens

Acer griseum
Acer palmatum
Amelanchier lamarckii
Cercis siliquastrum
Cornus kousa or *Cornus florida*
Crataegus laevigata
Malus domestica
Malus 'John Downie'
Prunus 'Pandora'
Sorbus hupehensis

I had a little nut tree, nothing would it bear
But a silver nutmeg and a golden pear.
The King of Spain's daughter came to visit me,
And all for the sake of my little nut tree.

Throwe al about your apple trees on the roots thereof, the urine of old men, or of stale pisse long kept, they shall bring fruit much better.

Leonard Mascall, 1592

Box is the hardest wood of any European tree – twice as hard as oak – and is said to be as durable as brass.

I think that a garden should never be large enough to be tiring, so that if a large space has to be dealt with, a greater part had better be laid out in a wood . . . I do not envy the owners of very large gardens.

Gertrude Jekyll (1843–1932)

The **London plane** is capable of surviving tough polluted conditions, such as those in the British capital when the pea-soup smog was at its most dense and toxic, partly because the bark renews itself by peeling off in strips. Many of the giant planes you see in Central London's squares and parks were planted over 200 years ago.

In an orchard there should be enough to eat, enough to lay up, enough to be stolen, and enough to rot on the ground.

James Boswell (1740–95)

The bark of the willow contains salicin, from which aspirin is derived.

THE **YEW** TREE, WHICH CAN live for up to 2,000 years, owes its fabled longevity to the very peculiar way that it grows. Branches root themselves and turn into trunks in their own right before joining up with the main trunk, which also explains the great width they reach. Don't sit in your garden and hold your breath for this to happen, however. It takes roughly 150 years for the main trunk of a yew to form. English archers, so feared by the French and their other enemies in the late Middle Ages, collected the wood for their longbows from churchyards up and down the country.

A **YEW** TREE IN FORTINGALL churchyard, near Loch Tay in Scotland, is believed to be between 3,000 and 5,000 years old.

THE **HAZEL**, DISTINGUISHABLE by its yellow catkins known as 'lamb's tails', is one of the most common trees in the British Isles, but growing to a maximum of 12 feet it is really more of a shrub than a tree. Its nuts, so cherished by squirrels, are the main ingredient in the much-maligned, sandals-and-socks dish the 'nut cutlet'. You can laugh at the nut cutlet until your sides need sewing up, but there is a very good reason why vegetarians hold it in such high esteem as a meat substitute and all-nutritious dish. Pound for pound, hazelnuts have five times more carbohydrates, six times more fat and twice as much protein as eggs. Put that in your pipe.

WHEN **LIFTING a tree** or shrub, always try to support it from underneath, either by holding the roots with your hands or, if the plant is too large, by placing it on some kind of covering like a polythene bag or some sackcloth and then carrying it (with the help of someone else) to its new location. There is a very sensible reason for this, namely that if you carry the young tree or shrub by the trunk or stem then the weight of the root ball will exert a heavy strain on the internal structure. By the same token, you would never pick up a baby or child by its neck.

A post of yew outlives a post of iron.

Old saying

HOMEMADE CHOCOLATE HAZELNUT SPREAD

If you are lucky enough to have a hazel tree in your garden, or near your home, your children will love you for ever if you convert the nuts into a fabulous chocolate hazelnut spread ✿ The nuts are usually baked on an oiled tray for about 10 minutes in a medium-hot oven, but you can fry them too ✿ Grind up the nuts, adding salt, butter and some melted dark chocolate, and mix into a paste.

Loveliest of trees, the cherry now
Is hung with bloom along the bough,
And stands about the woodland ride
Wearing white for Eastertide.

A. E. Housman
'A Shropshire Lad', 1887

Ceanothus shrubs originate from the Mediterranean-style habitat of the Californian chaparral or cowboy country. They have very shallow roots because they are accustomed to growing on dry, gritty slopes. It is advisable to stake them when you plant them so that they are not buffeted by strong winds. Ceanothus has a peculiar, dramatic habit of flowering itself to death after about 10 years. Blink and you'll miss the riot of colour and rapid death throes.

Hazelnuts are highly durable and archaeologists have found many dating back several millennia, petrified in ancient marshes and bogs. Our ancestors were grateful to the hazel for its tough but highly flexible young branches, which they used as fishing rods and for making baskets and fences.

As every schoolchild knows, **buddleias** attract butterflies like an overflowing honeypot attracts ants and bees. As many as 20 species are known to enjoy its sweet purple flowers. For the best new growth, you must be brutal and cut back a young plant in March to about two or three buds. It may feel cruel, but your buddleia and butterflies will thank you. With older plants it is best to proceed more gingerly, cutting back only about 30 or 40 per cent of the old growth.

MEDITERRANEAN PLANTS, SUCH as lavender, santolina, sage and rosemary, also enjoy a good short back and sides in early spring. This will prevent them getting woody and straggly, and ward off possible diseases.

BERRIES BECOME SWEETER AS THE months progress. **Gooseberries** are the first edible fruit of the year to appear in Britain followed by strawberries, raspberries, currants and finally blackberries. Gooseberries are also the tartest, and though some like the shock to the taste buds of eating them raw, most prefer to stew them and turn them into a fool simply by adding them to double cream with a bit of sugar and then leaving the mixture to chill in the fridge. The fruit is said to have got its name because it was once used to make a sauce with which to eat goose. The French do not venerate the gooseberry as we have done here over the years, but they enjoy it às a sauce with fresh mackerel, hence its name *'groseille à maquereau'*.

Trees are the best monuments that a man can erect to his own memory. They speak his praises without flattery, and they are blessings to children yet unborn.

LORD ORRERY, 1749

THE **MAGNOLIA** TREE, NAMED after the French botanist Pierre Magnol, is an ancient tree that evolved before the bees, and it developed in a such a way as to encourage its pollination by beetles.

Ash, mature or green,
makes a fire fit
for a Queen.

OLD SAYING

UP UNTIL THE END OF THE Second World War, the robust and fast-growing **alder** was prized for making charcoal. It was once used in the manufacture of gunpowder and its steady-burning properties made it handy in making fuses for explosives. The wood was also cherished for smoking fish and meat, and has been popular as a material for making electric guitar bodies.

ESTIMATES OF **BRITISH forest cover** through the ages:

500 BC 85%
AD 0 50%
1066 15%
1900 5%
2000 11.6%

THE WOOD OF THE **ASH** TREE IS remarkably strong and flexible, and is excellent for construction because it doesn't split when manipulated. In the past it was used to make stagecoaches, railway wagons, cartwheels, oars, early aeroplanes and the frame of the **Morris Minor** car.

A SINGLE edition of a daily newspaper uses paper made from the wood of approximately **5,000 trees**.

On an old 'cure' for hernias: *These* [ash] *trees, when young and flexible, were severed and held open by wedges, while ruptured children, stripped naked, were pushed through the apertures, under a persuasion that, by such a process, the poor babes would be cured of their infirmity.*

GILBERT WHITE
A Natural History of Selborne, 1789

OAK WOOD LASTS FOR CENTURIES because it doesn't rot, and its tightly packed grains make it almost rock hard. Our very early ancestors used it to make highly effective clubs for knocking animals and enemies senseless. In more recent times, its waterproof qualities made it perfect for the building of bridges and ships. The wood of the alder tree also refuses to decay in water, making it a prized material for the construction of bridges, jetties and boats, but it is soft and splits easily in dry conditions.

The [alder] *leaves gathered while the morning dew is on them and brought into a chamber troubled with fleas will gather them thereunto, which being suddenly cast out, will rid the chamber of these troublesome bedfellows.*

NICHOLAS CULPEPER
The Complete Herbal, 1653

THE **HAWTHORN** TREE OR SHRUB has a modest root system, which is why so many other plants are often to be found growing round it, grateful for the nutrients in the soil that other species of tree would have devoured. It is also why the hawthorn is so successful as part of a mixed hedgerow. The next time you are grazed by its thorny branches, stop before you curse this myth-laden tree. The flowers of the hawthorn contain a sedative and its medicinal properties have been welcomed since Victorian times, when a tincture made from the bright red berries was first prescribed as a remedy for various **heart conditions**, relieving mild to moderate angina, controlling blood pressure and increasing the heart's pumping power. Infusions made from the flowers, drunk several times a day, are said to help **angina**, and have also been taken for insomnia.

The best time to plant small trees is in mid-autumn, just before the leaves start to fall.

Hawthorn, an extremely common sight in our hedgerows and gardens and an important home for all sorts of wildlife, is regarded as providing the hottest firewood. It burns superbly even when green and makes first-rate charcoal that gets so hot it can melt pig-iron. It is thought that Christ's crown of thorns was made from hawthorn.

Pear wood is still used in the manufacture of high-quality woodwind instruments.

Before the introduction of metal to the manufacturing process, golf clubs were often made from apple wood, as well as beech, holly and pear.

One impulse from a vernal wood
May teach you more of man,
Of moral evil and of good,
Than all the ages can.

William Wordsworth
(1770–1850)
'The Tables Turned'

Hawthorn used to be called 'bread and cheese' because country people often added the young leaves to **sandwiches**. The young early-spring leaves of the beech tree and those of the lime tree make for excellent salads.

He who plants a tree loves others besides himself.

Old saying

Thrushes love the berries of the **rowan** tree and distribute the seeds in their droppings. Rowan used to be carried on ships to ward off storms and was also planted on graves to stop the dead returning to haunt the living.

The average tree will drink roughly 2,000 litres of water each year.

The **aspen** is believed to be the tree upon which Jesus Christ was **crucified**. As a result it was considered by many Christians to be cursed, which explains why its leaves tremble – supposedly in fear of its fate – at the slightest breath of wind.

WINTER

There are two seasons in Scotland – winter and July.

Billy Connolly

Every winter,
When the great sun has turned his face away,
The earth goes down into a vale of grief,
And fasts, and weeps, and shrouds herself in sables,
Leaving her wedding-garlands to decay –
Then leaps in spring to his returning kisses.

Charles Kingsley (1819–75)

If Winter comes, can Spring be far behind?

Percy Bysshe Shelley (1792–1822)

At Christmas I no more desire a rose
Than wish a snow in May's new-fangled mirth;
But like of each thing that in season grows.

William Shakespeare (1564–1616)

Perhaps I am a bear, or some hibernating animal underneath,
for the instinct to be half asleep all winter is so strong in me.

Anne Morrow Lindbergh (1906–2001)

Winter Birthdays

[?] December 1545: John Gerard, Nantwich, Cheshire
21 December 1918: Rosemary Verey, Chatham, Kent
30 December 1799: David Douglas, Scone, Perthshire
30 January 1913: Percy Thrower, Shropshire
13 February 1743: Sir Joseph Banks, London
15 February 1876: E. H. 'Chinese' Wilson, Chipping Campden, Gloucestershire

Be like the sun and meadow, which are not in the least concerned about the coming winter.

George Bernard Shaw
(1856–1950)

Winter vegetables

Cabbage, celery, Jerusalem artichokes, kale, leeks, oriental greens, parsnips, swede

France has neither winter nor summer nor morals. Apart from these drawbacks it is a fine country.

Mark Twain
(1835–1910)

Five flowers that flower in winter

Christmas rose (*Helleborus niger*)
Honeysuckle (*Lonicera fragrantissima*)
Mahonia (*Mahonia aquifolium*)
Viburnum 'Charles Lamont'
Winter jasmine (*Jasminium nudiflora*)

Every mile is two in winter.

George Herbert (1593–1633)

Children

There is a garden in every childhood

We may think that we are tending our garden, but of course, in many different ways, it is the garden and the plants that are nurturing us.

JENNY UGLOW
A Little History of British Gardening, 2005

IF **ALL THE WORLD'S WATER** WERE to fit into a large jug, the fresh water available for us to use would amount to no more than a teaspoon.

IF YOU WANT TO KNOW IF hedgehogs live in your garden or whether foxes pay you a visit at night, there is a simple way to find out: **make a footprint trap**. Buy a bag of soft sand and pour it into an area where the animals are most likely to appear. Make sure the sand surface is smooth and even. Sprinkle some fruit or vegetable pieces or some peanuts on the sand and then inspect for footprints the following morning. Make sure you don't use cooked food or you may find some horrible, dirty rats have got there first!

MAKE A LADYBIRD HOME. Ladybirds are gardeners' good friends as they eat a lot of unwelcome insects. It's simple to make a home for them to hibernate in over the winter. Take an old tin (e.g. one for chopped tomatoes or baked beans), wash it thoroughly and then pack it with wide drinking straws cut to the size of the tin. Towards the end of summer place it sideways in a hedgerow so that the rain doesn't get in, a few feet off the ground. The ladybirds, spiders and other insects will thank you for giving them a warm safe winter home.

IT TAKES ABOUT 27 LITRES OF water to grow a **single serving of lettuce**. Roughly 10,000 litres are required to produce one portion of steak.

Foods known or thought to have been introduced by the Romans

Almonds	Figs	Radishes
Apricots	Kale	Sour cherries
Artichokes	Leeks	Sweet chestnut
Asparagus	Marjoram	Turnips
Cucumbers	Onions	Walnuts
Dill	Parsley	
Fennel	Plums	

How much better, during a long and dreary winter, for daughters, and even sons, to assist, or attend their mother in a greenhouse than to be seated with her at cards, or in the blubberings over a stupid novel, or any other amusement.

William Cobbett
The English Gardener, 1829

Plant your very own conker or oak tree

This challenge obviously won't lead to instant gratification, but if nothing else, it teaches the virtue of patience. Collecting the acorns and conkers is a fun autumn activity in itself, but it is also a thrill to know that a small seed can turn into a tree towering over 100 feet high that could live for centuries. *Great oaks from little acorns grow* and all that . . .

Once you have collected the conkers or acorns, take a medium-sized flowerpot and place some broken crockery or large stones in the bottom before filling it up with a mixture of earth and compost. The seed should be planted about an inch deep in the middle of the pot. (It's a good idea to plant up a few pots in case some seeds don't sprout.)

Give them a good water before placing the pots in a semi-shady place in the garden. Check the pots occasionally over the winter to make sure the soil hasn't dried out. (The soil should be a little damp, but never really wet.) By the spring the seeds should have sprouted. Transfer the young trees to larger pots as they grow.

JOHN EVELYN, THE INFLUENTIAL gardener and garden writer of the mid-seventeenth century, leased his house, Sayes Court, in Deptford, to **Peter the Great** of Russia for three months in 1698 – and was horrified to discover the state of disrepair into which his garden had fallen when he returned. Peter apparently enjoyed being transported in a wheelbarrow in, around, through and over Evelyn's immaculately laid-out plots.

WORMS ARE ROUGHLY 1,000 times stronger than humans, relative to their size.

YOU MAY NOT BE ABLE TO tell by looking at them, but **earthworms** are actually very hairy. They are covered in thousands of tiny hairs which help them move through the earth.

The worm has turned

Children like worms far more than adults and are happy to let them wriggle around in their fingers without a care in the world. (Why is it that so many of us grow up to become a little squeamish about worms even when we know they are completely harmless? There are some people in the world who even eat them fresh out of the ground!)

One of the good things this project teaches, in these environmentally conscious times, is the value of recycling. You will be able to see this in action as the worms eat their way through your kitchen leftovers and turn it into an excellent compost.

Take a large clean glass jar, perhaps one that previously held some pickled cucumbers, and put about a finger's width of sand at the bottom. Add an inch or two of soil, and then some more sand and so on until you have almost reached the top (leave about two inches). Collect some earthworms by digging over a couple of patches of earth in the garden and put them into the jar with a few old leaves, vegetable peelings and pieces of fruit.

Put the lid on and puncture a couple of holes in the top for air. Worms hate light so tie some dark material or paper around the jar and put it away somewhere dark and cool for about two weeks. (They also hate being dry so make sure the soil is always pretty damp. It has been said that they like their

There was an Old Man with a beard,
Who said 'It is just as I feared!
Two Owls and a Hen, four Larks and a Wren,
Have all built their nests in my beard.'

EDWARD LEAR (1812–88)

environment to be about 35 per cent moisture.)

When you inspect the wormery and most of the vegetable matter has been pulled down into the soil, explain to your children why worms are so important to gardeners and farmers: worms aerate the soil, bring nutrients closer to the surface and create good drainage, and their poo (castings) is packed with goodies like potassium, phosphorus and nitrogen.

IN BRITAIN DOMESTIC **CATS KILL** up to 275 million animals, including 55 million birds every year, according to the Mammal Society. Many bird populations have been falling rapidly in recent years, largely as a result of modern farming practices, making it all the more important that we protect them as best we can. If you have bird feeders in your garden, make sure you hang them where cats cannot get at them, such as in or over prickly bushes and hedges, or from a metal pole or greenhouse that cats can't climb.

The human animal, in common with most other [species] *indigenous to our climate, is generally in high spirits and vigour during this month* [May]. *Woe to the young gardener who exhausts his spirits in any other way than self-improvement.*

J. C. LOUDON, 1830

THE LEAVES OF MANY PLANTS produce an oil to prevent them from **dehydrating** in the heat for the same reasons that we apply suntan lotion to our skin when the sun is at its hottest.

As the garden grows, so does the gardener.

OLD SAYING

OLDER CHILDREN CAN MAKE attractive, **natural necklaces** from seeds of pumpkins, melons or marrow and cherry stones. Wash the seeds well, leave them to dry in the sun and then lace them together using a long needle and some strong thread. You can dye the seeds or stones a lovely mahogany colour by soaking them in cold tea.

Ladybird, ladybird fly away home
Your house in on fire and your children are gone
All except one and that's little Ann
For she crept under the frying pan.

This popular children's song is thought to come from Kent, where the annual **burning of the hop plants** *killed off many ladybirds.*

I have never had so many good ideas day after day as when I worked in the garden.

JOHN ERSKINE (1879–1951)

AMAZE YOUR FRIENDS BY **GROWING your name** or **initials** on an apple or pear. Cut out the letters you want to create from some cardboard, making sure they are the right size for the fruit, and then paste them on to the side of the fruit at around the time it is due to change colour. Remove the letters when the fruit has ripened and you will be the proud owner of your very own trademarked apple or pear. You can perform a similar, quicker trick on your lawn by cutting out the letters from an old cereal box and staking them in the grass (or weighting them down with stones), preferably in a sunny part of the garden.

For a larger, more dramatic effect, which will *really* impress your friends, you can try this with a giant pumpkin, but instead of pasting on letters or patterns, carve them out on the skin by gently scraping the surface when the pumpkin is the size of a grapefruit. The patterns will grow at the same rate and size as the pumpkin itself, which will be that much larger at harvesting time if you remove the other fruits so that all the goodness is channelled into yours.

What pleasure have not children in applying their little green watering cans to plants in pots, or pouring water in at the root of favourite flowers in borders?

J. C. LOUDON
The Suburban Gardener and Villa Companion, 1838

Curious caterpillars

Collecting caterpillars and watching them turn into butterflies is an excellent, colourful and dramatic way of opening the eyes of children to the wonders of nature. When looking for caterpillars in the garden, keep an eye out for half-eaten leaves, which are often a telltale sign that they are around.

When you find a caterpillar, put it in a jam jar, but make sure you also put in the leaves of the plant they were eating, as most caterpillar species like only one or two plants. (If the caterpillars you find are hairy, pick them up using gloves or a leaf as the hairs can irritate the skin.)

When you have got them home, make sure the caterpillars have a regular supply of fresh leaves from the same plant. Don't worry about suffocating them when you put the lid on the jar because they will get all the air they need each day when you open the lid to feed them. Over the coming days, you will be able to watch each caterpillar get larger and then turn into a chrysalis or cocoon, which is the protective case in which the larva turns into an adult. They may look as if they have curled up and died, but don't worry, they're just steeling themselves for the moment they come of age. Soon enough, a butterfly will be born, and you can enjoy the thrill of releasing it into the wild.

A perfect summer day is when the sun is shining, the breeze is blowing, the birds are singing, and the lawn mower is broken.

JAMES DENT (N.D.)

QUEEN VICTORIA AND PRINCE Albert encouraged their children to garden. At Osborne House on the Isle of Wight they were each given their own vegetable and flower plot and their own personally monographed wheelbarrow and tools that had been scaled down to the right size.

A garden is like those pernicious machineries which catch a man's coat skirt or his hand, and draw in his arm, his leg, and his whole body to irresistible destruction.

RALPH WALDO EMERSON (1803–82)

Usually children spend more time in the garden than anybody else. It is where they learn about the world, because they can be in it unsupervised, yet protected. Some gardeners will remember from their own earliest recollections that no one sees the garden as vividly, or cares about it as passionately, as the child who grows up in it.

CAROL WILLIAMS
Bringing a Garden to Life, 1998

Nature trek

Taking your kids on a nature trek is a good way of getting them out of the house and enjoying the outdoors, while giving them – and you – a decent walk at the same time. You could even do a miniature version of it in your own garden if it's a reasonable size and has enough variety of plants and wildlife.

Before you set off, draw up a list of plants, insects, birds and animals to look out for. You can help them with the more difficult objects by showing them pictures in books before setting out. There are a number of other challenges to set them, such as finding flowers of different colours, leaves with X amount of points on them or objects with particular characteristics (flowers with a scent, plants with prickles or thorns, etc.). You could also play a form of I-Spy by sending them off to find plants that begin with a particular letter of the alphabet. If there are a number of children with you, you could add a competitive element by splitting them into teams, perhaps girls against boys. If they are going to wander out of your sight, be sure to warn them about potential hazards, such as not eating anything they find and not going too far away.

Bees are not as busy as we think they are. They just can't buzz any slower.

KIM HUBBARD (N.D.)

The kiss of the sun for pardon,
The song of the birds for mirth,
One is nearer God's heart in a garden
Than anywhere else on earth.

DOROTHY FRANCES GURNEY
(1858–1932)

TO SHOW CHILDREN HOW **plants need water** to survive and thrive, fill some glass jars with water and add roughly two dozen drops of food colouring. Take some flowers with white petals such as white carnations or daisies, and make sure you cut off the bottom of the stem at an angle to encourage water absorption. After a few hours the petals will start changing and after a day they will be a completely different colour. If you want to create an even whackier-looking flower, split the stem of a carnation down its length and place each half of it into separate jars of water mixed with different food colourings and see how the petals turn two different colours.

Gardening is a way of showing that you believe in tomorrow.

OLD SAYING

Six old sayings

Clear moon, frost soon.
Halo around the moon, rain soon.

If you see the underside of the
leaves in the gentle breeze,
It will rain before you sneeze.

When your joints all start to ache,
Rainy weather is at stake.

If February brings no rain,
'Tis neither good for grass nor
grain.

A Summer fog for fair,
A Winter fog for rain.

In October dung your field,
And your land its wealth will yield.

The gardening season officially begins on 1 January and ends on 31 December.

MARIE HUSTON (N.D.)

Plant your own vegetables

Many vegetables are very easy to grow and encouraging your children to grow some may even inspire them to eat them, rather than just push them round their plates, throw them against the wall or burst into tears at the sight of them. It's important that children have their own designated patches in the garden so that they feel a sense of ownership and stewardship. If you don't have enough room in the garden, growing vegetables in containers is a good alternative. An old barrel or tub would be fine and you can even grow some of the smaller varieties in window boxes or hanging baskets so long as you are vigilant about watering. Using an old carriage wheel is a colourful and practical way for kids to start growing their own, with the spokes of the wheel neatly dividing the area into obvious sections. Painting the wheel in their favourite colours before laying it on the ground would be a fun activity in itself.

In order that the children can see some quick results from their hard work it's probably a good idea to include some fast-growing vegetables like salad leaves, radishes, spring onions,

10 easy-to-grow vegetables

Beetroot
French beans
Lettuce
Potatoes
Radish
Rhubarb *(yes, it's a vegetable!)*
Rocket
Runner beans
Spinach
Swiss chard

spinach, chard and rocket, which have the additional virtue of being pretty unfussy. Baby carrots (which most children seem to enjoy eating and which don't take up too much space) are also a good option. Tomatoes are another favourite, but they involve a bit more care and for guaranteed results you're better off growing them under glass in pots or grow-bags.

You will obviously have to supervise your children in preparing the bed or container, explaining to them the benefits of composting, planting at the right depth, watering and weeding and so on. If they are planting very fine seed which can easily end up in lots of clusters, show them how to spread it evenly by placing the seed in an envelope with a small corner cut off so that they can just tap it into the soil without it all pouring out. It would help if the children had their own small watering cans, partly because the gallon-sized ones will be too heavy for them, but also because it will make them feel even more dedicated to their project. The seeds should be given a light watering after planting and once the plants start to emerge you should encourage the children to check for signs of slug damage, and make a trap for the pests by submerging a plastic jar filled with beer into the soil. You might also want to plant or pot up some marigolds and nasturtiums to ward off aphids and create a bit of colour. If you've got the room, it might be worthwhile growing something a little more dramatic, such as pumpkins, courgettes or cucumbers. The pumpkins will be a sure-fire success because they can be used at Hallowe'en to make jack-o'-lanterns. They also grow so fast and so copiously that the children will notice fresh growth almost by the day.

10 easy-to-grow flowers

Campanula
Forget-me-not
Geranium
Lavender
Marigold
Nasturtium
Pansy
Primrose
Sweet pea
Wild strawberry

There is a garden in every childhood, an enchanted place, where colours are brighter, the air softer, and the morning more fragrant than ever again.

ELIZABETH LAWRENCE (1904–85)

MAKING **BUTTERFLY PAINTINGS** is fun and easy for younger children to do on a rainy day. Get them to fold a piece of A4 paper in half and then paint a butterfly, or half of one, on one side of the paper. Fold the paper over immediately while the paint is wet and then open it up to reveal either two butterflies or a whole one.

Recreate the desert in your bedroom

The curious shapes and exotic nature of the cacti family have always held a fascination for young children. The notion that they can grow these prickly, weird-looking plants that they never see by the roadside or in the gardens of Britain is part of the appeal. It might also have something to do with the fact that they are so simple to look after. Once they have been planted up, cacti need only an occasional watering and will happily survive months of being ignored. Creating your own mini-desert environment is fairly straightforward too, and a wide range of cactus plants is available at most garden centres.

For a larger 'desert' you will need an old washing-up bowl or big flowerpot with holes punctured in the bottom for drainage, but an ordinary medium-sized flowerpot or a

small tub of some sort is fine if your plans are less grandiose. Either way, place some broken pieces of a terracotta pot in the bottom of the container (your children will enjoy being invited to smash a pot) and then cover them with some grit and fill the rest of the pot with a mixture of coarse sand and compost. When planting the cacti, use a spoon to dig out a hole and make sure the soil around the base of each plant is well patted down. Some cacti have furry prickles or fairly smooth skins, but when planting up the prickly ones, use a garden glove or a cloth. (Getting a tiny thorn out from under the skin can be more difficult and painful than removing a splinter.) Once all the plants have been potted, sprinkle a mixture of sand and small stones over the surface to create a desert effect.

Cacti, needless to say, love the sun, so make sure you place the mini-desert on a south-facing windowsill. The plants don't need much watering but when the soil beneath the stony surface has completely dried out, use a small watering can or teapot to give them a good drink.

Everybody talks about the weather but nobody does anything about it.

Charles Dudley Warner (1829–1900)

Children can cut out the front and back of seed packets of plants they have sown and stick them on to pages; they can press small flowers and leaves and glue them in; make a note of the different animals, birds and insects they have spotted in the garden and then keep a running count of their sightings. (Who is the most frequent visitor? Will, for instance, the blackbird knock the ubiquitous robin off his proverbial perch?) One good way of drawing attention to

Grow your own scrapbook

Once children have set, sown and potted their plants, a good way to maintain their interest in the garden is to encourage them to make their own scrapbooks in which they can record all the changing events of the season. The scrapbook also has the virtue of giving them something to do on a rainy day spent indoors. You can apply your own imagination as to how to go about this, but there are a number of simple exercises you may want to try.

the changing nature of the seasons is to take photographs at regular intervals (perhaps every two weeks) of certain plants in the garden, recording how they emerge from the soil, come into leaf, blossom, fruit and then die away. If you're really stuck for options and have exhausted all other activities and the rain is still coming down, there is always the fallback of getting children to sketch or paint a particular view of the garden, or a favourite plant!

Bat, bat, come under my hat,
And I'll give you a slice of bacon;
And when I bake,
I'll give you a cake,
If I am not mistaken.

ANON

THE MOST COMMON VARIETY OF bat to be found in the British garden is the small **pipistrelle**, which does no harm to people or property, and likes to roost in the dry hollows of trees, behind tiles or in the small crevices of buildings. If there is no obvious habitat for bats in your garden, you can build a simple bat box or buy one from the RSPB or from a garden centre. Many websites carry simple instructions as to how to build a box. Make the box from rough-sawn timber to give the bats something to cling to, and be sure the wood is untreated as many preservatives kill bats. The best spot for a box is under the eaves of a house or on a tree, as high as possible above the ground to avoid predators. Remove surrounding branches to give the bats a clear flight path.

THE **MUCH MISUNDERSTOOD bat** has long suffered a bad press, but its usefulness in the garden as a predator of many unwelcome insects is being recognized again. Bat numbers have declined dramatically in recent years, partly because the use of chemicals in farming has killed off so many insects on which they feed, and partly because their habitats have been gradually disappearing. Bats are now well protected by the law and you would be committing a criminal offence if you injured one or interfered with its home even if it was on your property.

I was much entertained last summer with a tame bat, which would take flies out of a person's hand . . . The adroitness it showed in shearing off the wings of flies, which were always rejected, was worthy of observation and pleased me much.

GILBERT WHITE
Natural History of Selborne, 1789

MAKING APPLE DOLLS IS A traditional American pasttime stretching back to the days of the early settlers in the 1600s and 1700s, when people had to make their own fun and their own toys with the limited amount of materials available to them. Making apple dolls encourages children to exercise their imaginations by creating a whole host of characters. They can make silly faces or monsters or try to produce likenesses of their family and friends.

20 poisonous plants in your garden

Busy Lizzie (young stems, leaves)
Buttercup (leaves)
Clematis (leaves)
Daphne (leaves, berries)
Elder (shoots, leaves)
Foxglove (leaves)
Hellebore (leaves)
Hydrangea (leaves)
Laburnum (leaves, seeds)
Laurel (all)
Lily of the valley (leaves, flowers)
Lobelia (leaves, stems)
Mistletoe (berries)
Monkshood (all)
Poinsettia (leaves)
Primrose (leaves, stems)
Privet (leaves, berries)
Rhubarb (leaves)
Wisteria (all)
Yew (all)

Peel an apple and quickly rub some diluted vinegar or lemon juice over the surface to stop the oxygen from turning it brown. Carve a face into the apple using a potato peeler or any other appropriate and safe kitchen utensil. Dab some more vinegar or lemon into the features you have cut out and then roll the doll in some table salt on a plate to dry it out. Thread some firm, strong wire through the centre of the core before hanging the doll up to dry somewhere warm like in an airing cupboard, but keep it out of direct sunlight. You can dry the dolls on a tray too, but be sure not to handle them while they dry. After two or three weeks the dolls should be dry, and this is when the fun bit starts. If you're feeling adventurous you can make a body for the doll, by tying wires or pipe cleaners into the shapes of limbs and dressing the figure in doll's clothes, and you could also buy some styrofoam cones and paint them. You can paint the faces (or use colourful beads for the eyes) and use wool or cotton for the hair.

Turn your mum or dad into a scarecrow

Birds can be a major problem if you're growing vegetables or fruits in your garden, but you can borrow a trick that farmers have been using since man first started growing his own food: make a scarecrow to frighten them off. For the main body of the scarecrow, you need to make the shape of a cross, either by nailing one thin wooden board to another or by tying together two poles or thick sticks. Then dig out some of your mum and dad's old clothes that they were planning to discard and dress up the figure in whichever way you choose. For the head, stuff an old T-shirt with cloth and hold it together with safety pins or rubber bands and sow on some large buttons for the eyes (or draw them on with a felt tip if the material is a light colour). For the nose, take a pointy carrot, make a small hole where it should go and slot it through before you seal the head. If your mum or dad has a big nose, use a big carrot.

THE FRENCH WORD FOR dandelion is *pissenlit*, which means 'wee in bed'. This is because the leaves of the plant are diuretic and help increase the formation and flow of our pee.

A modest garden contains for those who know how to look and wait, more instruction than a library.

HENRI FRÉDÉRIC AMIEL (1821–81)
Swiss philosopher

THE LONG STRINGS OF GOLDEN flowers on the **laburnum tree** are beautiful but the seeds, which are attractive to kids because they look like peas in a pod, are deadly poisonous. Symptoms of poisoning are intense sleepiness, vomiting, spasms and dilated pupils. If you have a laburnum tree in your garden, be sure to warn your children of the dangers and make sure very small toddlers or babies are not allowed to crawl or play near it once it is in seed. Doctors advise that parents do not try to induce vomiting if a child has eaten the seed, but instead immediately call for medical help.

Hairy eggheads

When you cook some eggs, try to crack them open so that only the very top of the shell comes off, leaving it as close to its original size as possible. Wash the shells and let them dry before painting a face on each or drawing one with a felt-tip pen. When this is done, fill the shells with some compost and sprinkle on some grass seed, which you cover with a little more compost before firming down and pouring a small amount of water over the surface. Put the eggheads back into the carton and make sure the soil remains a little damp – not wet – while you wait for the grass to grow. Soon enough, you should have some ridiculous-looking eggheads with hair growing straight up as if they have received an electric shock, or seen something highly alarming like their dad mowing the lawn in a bikini. Once the grass has reached a good height, you can give your character any hairstyle you like – a Mohican or a mullet for instance – by cutting the grass to shape with a small pair of scissors.

MOTHS AND MOTH caterpillars provide a valuable source of food for birds and bats. There are about 2,500 species in Britain alone. Many **moths** are as beautiful as butterflies and because they are drawn to bright lights, it's very easy to attract them to your house to observe them close up. If you have an outside light, attach all four corners of an old thin pillowcase, or muslin cloth, to the bottom of the light and when it starts to get dark you will soon have a whole squadron of moths zooming in on the material. This is perfect for moths because much as they like the light of your lamp, they can't settle on it because the heat of the light bulb is too great. The contraption you have made will give them the best of both worlds while allowing you to have a good look at them.

On bees: *What two extraordinary substances to be made, by little creatures, out of roses and lilies! With singular and lively energy in nature to impel these little creatures thus to fetch out the sweet and elegant properties of the coloured fragrances of the garden, and serve them up to us for food and light – honey to eat, and wax papers to eat it by!*

LEIGH HUNT
(1784–1859)

Index